THE SIGNIFICANCE OF A NON-REDUCTIONIST
ONTOLOGY FOR THE DISCIPLINES OF

MATHEMATICS
& PHYSICS

A HISTORICAL AND SYSTEMATIC ANALYSIS

DANIËL F.M. STRAUSS

www.paideiapress.ca
www.reformationaldl.org

Mathematics & Physics
A publication of Paideia Press (3248 Twenty First St., Jordan Station, Ontario, Canada L0R 1S0).

© 2021 by Paideia Press. All rights reserved.

All rights reserved. Except for brief quotations in critical publications or reviews, no part of this book may be reproduced in any manner without prior written permission from Paideia Press at the address above.

Cover Design by Steven R. Martins

Typeset by D.F.M. Strauss

ISBN 978-0-88815-301-2

Printed in the United States of America

Contents

Overall perspective

.... 1
The contemporary intellectual climate 1

I
Mathematics

.... 3
Are there different standpoints in mathematics? 3
Two apparently simple questions with
'self-evident' answers 4
Historical detour 4
Starting-points for a third alternative? 6
Numerical and spatial addition in the context
of the law-subject distinction. 6
Distance: highlighting the mutual coherence
between number and space 10
Back to space 12
What is presupposed in space? 13
What is the interrelation between space and number? 14
    Which region is more basic? *14*
    Interconnections between functional domains *17*
The irreducible meaning of space underlying
Hilbert's primitive terms. 18
The theory of modal aspects 21
The impasse of arithmeticism 23
The circularity entailed in set theoretical attempts to
arithmetize continuity 29

II
Physics

.... 33
Historical perspective on the concept of matter 34
The mystery of matter. 44

i

The problem of individuality 48
Systematic distinctions 50
    Knowledge based upon universality *50*
    Knowledge exceeding universality *51*
    The concept of a normative principle and a natural law *53*
Physical entities exceed the limits of physics 58
Concluding remark 62

Appendix I
The 'exactness' of a mathematical proof dependent upon philosophical assumptions 63
Non-denumerability: Cantor's Diagonal Proof 63

Appendix II
The wave-particle duality 66
Complementarity – limits to experimentation 66

The wave particle duality and the idea of the typical totality structure of an entity 66

Bibliography
.... 68

Index of Subjects
.... 73

Index of Names
.... 77

Overall perspective[1]

The attempt to reduce what is truly unique to something else leads to the deification of something or some aspect within creation, normally accompanied by imperialistic "all"-claims such as, "everything is number," "everything is matter," "everything is feeling," "everything is historical" or "everything is interpretation." The distortions thus created inevitably result in insoluble *antinomies*.

A Christian approach to scholarship, directed by the central biblical motive of creation, fall and redemption and guided by the theoretical idea that God subjected all of creation to His Law-Word, delimiting and determining the cohering diversity we experience within reality, in principle safe-guards those in the grip of this ultimate commitment and theoretical orientation from *absolutizing* anything within creation.

1 The contemporary intellectual climate

In spite of the decline of positivism within the domain of the *philosophy of science* many scholars in the various academic disciplines (special sciences) still advocate its *neutrality postulate*. As examples of "exact sciences" mathematics and physics are normally lifted out. These two disciplines, according to the positivistic view, are *objective* and *neutral* – they rule out the possibility of any and all presuppositions exceeding the boundaries of these "exact" disciplines. Alternatively, insofar as *historicism* and its relativistic consequences gave rise to what is known as *postmodernism*, grand stories ("meta-narratives") are questioned and truth uprooted – every person has her own "story" to tell.

Amidst all of this the leading philosophers of science during the 20th century increasingly acknowledged the inevitability of an ultimate commitment in scholarship as well as the presence of a (philosophical) theoretical view of reality underlying all academic endeavours. Some of them explicitly reject *reductionism*. Popper straightforwardly states: "As a philosophy, reductionism is a failure" (Popper, 1974:269). In order to capture problematic situations within the disciplines (and their logic) the term *reductionism* emerged by the middle of the 20th century. In 1953 Quine used it in his discussion of "The Verification Theory and Reductionism" (see Quine, 1953:37 ff.) and in the early seventies the work "Beyond Reductionism" appeared (see Koestler & Smythies, 1972). Smith (1994) considers the scientist-philosopher Michael Polanyi to be "perhaps the severest and most comprehensive critic of reductionism" because he "was a major scientist of this century and was drawn into philosophical debate primarily because of the threat to scientific

---

1 Paper presented at the *Metanexus Institute* in Madrid – General Conference Theme: *Subject, Self, and Soul: Transdisciplinary Approaches to Personhood.* (July 2008 – by *Prof D F M Strauss*: dfms@cknet.co.za).

freedom, political democracy, and to humane values that he saw in reductionism". To this he adds the remark:

> His works *The Contempt of Freedom*, *The Logic of Liberty*, *Science Faith and Society*, *Personal Knowledge*, and *The Tacit Dimension* have as a common theme the criticism of reductionism in all its scientific, cultural and moral forms.

The best way to challenge both positivism (objectivity and neutrality) and postmodernism (historicism and relativism) is to confront the supposedly "exact" sciences, mathematics and physics, with the implications of a non-reductionist Christian philosophy.

I
Mathematics

Practicing mathematicians, consciously or not, subscribe to some philosophy of mathematics (if unstudied, it is usually inconsistent) (Monk, 1970:707)

2 Are there different standpoints in mathematics?

Before we investigate relevant historical perspectives and systematic distinctions it is worth challenging this claim of Fern by quoting a number of statements:

The mathematician Kline writes:
> The developments in the foundations of mathematics since 1900 are bewildering, and the present state of mathematics is anomalous and deplorable. The light of truth no longer illuminates the road to follow. In place of the unique, universally admired and universally accepted body of mathematics whose proofs, though sometimes requiring emendation, were regarded as the acme of sound reasoning, we now have conflicting approaches to mathematics (Kline, 1980:275-276)

In respect of formalization in intuitionistic mathematics the Dutch logician Beth remarks:
> Meanwhile, for the intuitionists this formalization has in no way the meaning of a foundation as it does for the logicists. On the contrary, formalistic expression is in a position to produce no more than an inadequate picture of intuitionism (Beth, 1965:90).

The intuitionistic mathematician, Heyting, explains what is basic to intuitionism
> every logical theorem ... is but a mathematical theorem of extreme generality; that is to say, logic is a part of mathematics, and can by no means serve as a foundation for it (Heyting, 1971:6).

Of course intuitionism represents an authentic mathematical stance:
> The intuitionists have created a whole new mathematics, including a theory of the continuum and a set theory. This mathematics employs concepts and makes distinctions not found in the classical mathematics (Kleene, 1952:52).

In fact intuitionism created an entirely *new mathematics*. Beth explains:
> It is clear that intuitionistic mathematics is not merely that part of classical mathematics which would remain if one removed certain methods not acceptable to the intuitionists. On the contrary, intuitionistic mathematics replaces those methods by other ones that lead to results which find no counterpart in classical mathematics (Beth, 1965:89).

Perhaps the most perplexing observation comes from Stegmüller:
> The special character of intuitionistic mathematics is expressed in a series of theorems that contradict the classical results. For instance, while in classical mathematics only a small part of the real functions are uniformly continuous,

in intuitionistic mathematics the principle holds that any function that is definable at all is uniformly continuous" (Stegmüller, 1970:331).

3 Two apparently simple questions with 'self-evident' answers

(i) Is 2 + 2 = 4? and (ii) is a straight line the shortest distance between two points?

We may relate question (i) to the idea that mathematics is objective and neutral – as asserted by Fern:

> Mathematical calculations are paradigmatic instances of universally accessible, rationally compelling argument. Anyone who fails to see "two plus two equals four" denies the Pythagorean Theorem, or dismisses as nonsense the esoterics of infinitesimal calculus forfeits the crown of rationality (Fern, 2002:96-97).

The statement that "a straight is line the shortest distance between two points" indeed seems to be as self-evident as the statement that "2 + 2 = 4". In an earlier phase of his development Bertrand Russell 'corrected' this definition: "A straight line, then, is not the shortest distance, but is simply the distance between two points" (Russell, 1897:18). The three key terms in this statement concern *spatial* configurations, namely the terms 'line', 'point' and 'shortest'. Yet the crucial element maintained in Russell's improved definition echoes something of our awareness of numerical relations: *distance*.[1] If this is indeed the case it may turn out that an analysis of this statement will at once get entangled in the consideration of arithmetical and spatial issues, which means that it cannot be analyzed purely in *spatial* (or *geometrical*) terms.

4 Historical detour

Early Greek mathematics followed the arithmeticistic approach of the Pythagorean school with its claim that "everything is number." Although the Pythagoreans believed that numerical relationships ordered the cosmos, they discovered that geometrical figures and lines can be construed that cannot be expressed by the relation between integers. The discovery of incommensurability by Hippasus of Metapont (450 B.C.) therefore caused a crisis since within the assumed form-giving function of number the formless (infinite) was revealed. Laugwitz remarks: "Every numerical relationship allows for a geometric representation, but not every line-relationship can be expressed numerically. This established the primacy of geometry over arithmetic and as a result the Books of Euclid treat the theory of numbers as a part of geometry"

---

[1] The focus of our considerations will be on the interconnections between space and number. We shall argue that there are structural features that are inherent in these two facets of reality prior to the actual definition of *metrical* spaces (in 1906 by Fréchet). Mac Lane accepts space as "something extended" and on the basis of the notion of 'distance' defines a *metric space* (see Mac Lane, 1986:16-17). It is clear that the notions of *extension* and *distance* **precede** the definition of a metrical space. An explanation of the mutual relation between discreteness and continuity within a *topological* context requires a different argument. A starting-point for such a discussion is found in White (1988:1-12).

(Laugwitz, 1986:9).[1] This geometrization of mathematics inspired a space metaphysics lasting at least until Descartes and Kant. During the nineteenth century, however, Cauchy, Weierstrass, Dedekind and Cantor once again pursued the path of an *arithmeticistic* approach. Of particular significance in this regard is set theory as it was developed by Cantor (including his theory of *transfinite arithmetic*).

When Russell and Zermelo independently discovered the fundamental inconsistency of Cantor's set theory in 1900 and 1901, mathematics gave birth to three schools of thought, namely the *logicist* school (Russell, Gödel), the *intuitionist* school (Poincaré, Brouwer, Heyting, Weyl and Dummett), as well as the *axiomatic formalist* school (guided by the foremost mathematician of the twentieth century, David Hilbert and still largely dominating the scene of contemporary mathematics).

In 1781, the first edition of his influential work, "Critique of Pure Reason" (CPR) appeared. What is remarkable is that the main systematic subdivisions of this work provide the springboard for the three mentioned diverging trends in 20th century mathematics – intuitionism (exploring the "transcendental aesthetic" of the CPR), logicism (oriented to the "transcendental analytic") and axiomatic formalism (affirming the thrust of the "transcendental dialectic"). We already noted that the fundamental differences within the discipline of mathematics caused a situation where what is true within intuitionistic mathematics may be false within formalism, while what is mathematically accepted by formalism, such as Cantor's theory of transfinite numbers, is rejected as a phantasm by intuitionism (see Heyting, 1949:4) and as non-existent.[2]

In 1900 the French mathematician, Poincaré, made the proud claim that mathematics has reached absolute rigour. In a standard work on the foundations of set theory, however, we read: "ironically enough, at the very same time that Poincaré made his proud claim, it has already turned out that the theory of the infinite systems of integers – nothing else but part of set theory – was very far from having obtained absolute security of foundations. More than the mere appearance of antinomies in the basis of set theory, and thereby of analysis, it is the fact that the various attempts to overcome these antinomies, ..., revealed a far-going and surprising divergence of opinions and conceptions on the most fundamental mathematical notions, such as set and number themselves, which induces us to speak of the third foundational crisis that mathematics is still undergoing" (Fraenkel, A. et al, 1973:14).

In this context the history of Gotllob Frege is perhaps the most striking. In 1884 he published a work on the foundations of arithmetic. After his first Volume on the basic laws of arithmetic appeared in 1893 Russell's discovery (in 1900) of the antinomous character of Cantor's set theory for some time delayed the publication of the second Volume in 1903 – where he had to concede

---
1   In connection with the history of the concept of matter we shall return to Greek philosophy.
2   Just compare the remarks quoted above concerning different standpoints in athematics.

in the first sentence of the appendix that one of the corner stones of his approach had been shaken. Russell considered the set C with sets as elements, namely all those sets A that do not contain themselves as an element. It turned out that if C is an element of itself it must conform to the condition for being an element, which stipulates that it cannot be an element of itself. Conversely, if C is not an element of itself, it obeys the condition for being an element of itself.

Close to the end of his life, in 1924/25, Frege not only reverted to a geometrical source of knowledge, but also explicitly rejected his initial logicist position. In a sense he completed the circle – analogous to what happened in Greek mathematics after the discovery of irrational numbers. In the case of Greek mathematics this discovery prompted the geometrization of their mathematics, and in the case of Frege the discovery of the untenability of his "Grundlagen" also inspired him to hold that mathematics as a whole actually is geometry:

> So an a priori mode of cognition must be involved here. But this cognition does not have to flow from purely logical principles, as I originally assumed. There is the further possibility that it has a geometrical source. ... The more I have thought the matter over, the more convinced I have become that arithmetic and geometry have developed on the same basis – a geometrical one in fact – so that mathematics in its entirety is really geometry (Frege, 1979: 277).

What is therefore the upshot of the history of mathematics? It emerged under the spell of Pythagorean arithmeticism ("everything is number"), then, owing to the discovery of irrational numbers (incommensurability) it experienced a fundamental geomatrization and during the 19th century it once again explored the avenue of arithmeticism, thus closing the circle of arithmeticism. Finally, in the thought of Frege also the circle of space was closed, because he once more thought that all of mathematics fundamentally is geometry.

5   Starting-points for a third alternative?

From the perspective of a non-reductionist ontology there is an obvious alternative never pursued throughout the history of mathematics:

> Accept the uniqueness and irreducibility of number and space as well as their mutual interconnectedness and coherence.

In order to highlight this alternative way we may use the above-mentioned argument regarding $2 + 2 = 4$ as our angle of approach. This will introduce considerations stemming both from the domains of number and space, particularly through the introduction of another 'sum' – one in which we may suggest that "2+2" is not equal to 4 but equals $\sqrt{8}$.

6   Numerical and spatial addition in the context of the law-subject distinction

The conviction that mathematics is *objective* and *neutral* may here be defended by reinforcing the original claim (namely that 2+2=4), i.e. by referring to the addition of 2 fingers and another two 2 fingers, which indeed adds up to

4 fingers. Apparently this specified addition conclusively confirms the soundness of the initial statement that 2+2 is equal to 4. Unfortunately the issue is more complicated than it may seem at first sight, for the alternative assertion, namely that 2+2=√8, implicitly changed the *context of addition*. When a person walks 2 miles north and afterwards 2 miles east, then that person will be √8 miles away from the initial point of departure. This context concerns an instance of *spatial addition* that is mathematically treated in *vector analysis*, where a vector possesses both *distance* (magnitude) and *direction*.[1] One can capture this altered context by underscoring the numerals involved in order to specify the fact that we are dealing with vectors: 2+2=√8. The upshot is that we now clearly have two different *kinds of facts* related to addition at hand: a *numerical* fact (designated as 2+2=4) and a *geometrical* fact (in the "right-angle-case" designated as 2+2=√8). In order to capture the specifications of this example one may construct the following figure:

These facts are not *unqualified* – that is to say, they are *distinct* because they are differently qualified or structured, respectively as *numerical* and as *spatial* facts. They are therefore not simply 'facts' in themselves. In their factuality they are *delimited* by alternative order-determinations. The operation of numerical addition displays an order-determination different from the operation of spatial addition, as is clearly manifested in the alternative sums: 4 and √8. In our example the underlying "order diversity" therefore makes possible the indicated distinction between numerical and spatial facts.

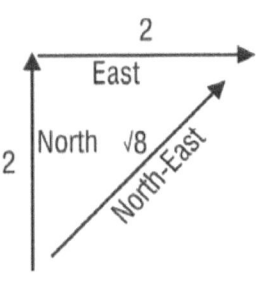

But there is something else present in this distinction between these two kinds of facts, namely the reference to the *operation of addition*. Modern mathematical set theory normally first of all approaches this domain in terms

---

1   In the first half of the 19th century Grassmann already introduced the idea of a vector. He designnnated such a line segment ('Strecke') with a specific direction and length on a specific straight line as a "linienteil" which became more generally known in German literature as a *vector* ('Vektor'): "Graßmann nannte eine solche Strecke bestimmter Richtung und Länge auf einer bestimmten Geraden einen *Linienteil*; jetzt ist in der Deutschen Literatur der Name Vektor üblicher" (Klein, 1925:24). Hedrick and Noble mistranslated "Richtung und Länge" as "length and sense," perhaps because later on in the same original German paragraph Klein himself used the German words "Länge und Sinn") – also explaining why the word order of "Richtung und Länge" was reversed to "length and sense" by them (see Klein, 1939:22). Please note that our example (2+2=√8) merely refers to one instance of a vector sum – the case where the connecting angle is 90°. By varying the angle the sum could take on any value between (and including) 2 and 4.

of the algebraic structure of *fields* – where the (binary) operations called addition (+) and multiplication (.) meet the field axioms (specified as *laws*).[1]

Let us give one step back and initially extract from this mathematical practice merely the operations (laws) of addition and multiplication. The fact that addition and multiplication within a specific system of numbers (such as the *system* of natural numbers) yield numbers still belonging to the initial set is also mathematically articulated by saying that the *system of numbers* under consideration is *closed* under the operations (laws) of *addition* and *multiplication*. Since ancient Greek philosophy it was understood that conditions (laws) and whatever meets these conditions are both distinct and strictly correlated. The most basic instance of such a strict correlation between (arithmetical) laws and arithmetical subjects (numbers) conditioned by these laws is found in the system of natural numbers. It is immediately evident that the addition and multiplication of any two natural numbers once more yield natural numbers (s = system; t = set):

system of na- tural numbers N_s	operations / laws: numerical subjects:	$(+,\times)$ $N_t = (1, 2, 3, ...)$

The designation 'system' therefore comprises both *arithmetical laws* and *arithmetical subjects* – in the sense that the laws (operations) not only *determine* the behavior of the subjects but also *delimit* them. What has been explained above therefore means that the system of natural numbers finds its *determination* and *delimitation* in the operations of addition and multiplication that are closed over the set of natural numbers – in the sense that adding or multiplying any two natural numbers will always yield another natural number. The ultimate presupposition of these operations is found in the numerical order of succession. The Peano axioms (for the positive integers) yield a mathematical articulation of this primitive arithmetical order of succession. The correlation of the operations of addition and multiplication and their delimiting and determining role in respect of numerical subjects are consistent with Peano's axioms because they are entailed in the *complete ordered field of real numbers* (see Berberian, 1994:230).

Introducing further arithmetical laws or operations will invariably call for additional (correlated) numbers that are factually subjected to the determining and delimiting arithmetical laws. For example, if the operation of subtraction is added to those of addition and multiplication, the correlating set of integers (Z_t) is constituted – and considered in their correlation this yields the *system of integers* (see Ebbinghaus, et.al, 1995:19).

system of integers I_s	-	operations / laws: numerical subjects:	$(+,\times,-)$ $I_t = (0, +1, -1, +2, -2, ...)$

---

[1] A *field* is defined as a set F such that for every pair of elements a, b the sum $a+b$ and the product ab are still elements of F subject to the associative and commutative laws for addition and multiplication, and combined to the presence of a *zero element* and a *unit* (or *identity*) element (see Bartle, 1964:28; Berberian, 1994:1 ff.). This definition of a field is then expanded to that of an ordered field and it is finally connected to the idea of completeness.

Likewise, extending the arithmetical operations by introducing *division* the correlating *set* of fractions is needed within the *system* of rational numbers.

system of rational numbers Q_s	operations / laws: numerical subjects:	$(+, \times, -, :)$ $Q_t = (a/b; a,b \in Z_t / b \neq 0)$

This explanation, in terms of the strict correlation between operations at the law-side and numerical subjects at the factual side, is *formally* similar to the way in which Klein introduces negative numbers and fractions (by means of the reverse operations of addition and multiplication – see Klein, 1932:23 ff. & 29 ff.). Ebbinghaus *et.al* points out that in a Paper on "Pure Number Theory" (*Reine Zahlenlehre*) Bolzano already developed a theory of rational numbers, "and in fact a theory of those sets of numbers that are *closed with respect to the four elementary arithmetic operations*" (Ebbinghaus, 1995:22).

Against this background it is clear that the systematic arithmetical statement 2+2=4 does not designate a "brute fact" (a fact "in itself," "an sich"), since the factual relation specified for numerical subjects (selected from the set of natural numbers) that are involved in it, exhibits the *measure* of the numerical law of addition. One can also say that this statement conforms to the determining and delimiting effect of the arithmetical law of addition. Consequently, the statement that 2+2 is equal to 4 concerns a law-conformative (arithmetical) state of affairs – it displays a specific lawfulness or orderliness for it meets the conditions set by the presupposed arithmetical order.

If there are multiple laws known to be *arithmetical laws* then one may speak of a unique *sphere of arithmetical laws* strictly correlated with diverse arithmetical subjects (sets of numbers) subjected to these laws. Another way to capture this situation is to speak of a numerical sphere in which arithmetical laws are strictly correlated with arithmetical subjects (numbers); in other words within this numerical domain a distinction is made between its *law-side* (order side) and its *factual side*. Myhill, who appreciates Brouwer as the originator of "constructive mathematics," introduces the notion of a 'rule' (the equivalent of what we have designnated as "law-side") as "a primitive one in constructive mathematics"; "We therefore take the notion of a rule as an undefined one" (Myhill, 1972:748). (Myhill received his Harvard Ph.D. under W.V. Quine). In his encompassing introduction to set theory (the third impression), Adolf Fraenkel refers to the peculiar *constructive* definition of a set which accepts as a foundation the *concept of law* and the *concept of natural number* as *intuitively given* (Fraenkel, 1928:237).[1]

---

[1] "Ohne die Stellung wieterer intuitionistischer Gruppen und anderer Richtungen ... zum Mengenbegriff zu schildern, sei hier noch auf die wesentlich abweichende Auffassung Brouwers hingewiesen. Dieser stellt eine eigenartige rein *konstruktive* Mengendefinition an der Spitze, bei der der Begriff der *natürliche Zahl* und der des *Gesetzes* als intuitiv gegeben zugrunde gelegt werden."

The geometrical sum – 2+2=√8 belongs to a different domain, to a different sphere of laws, one where it is also possible to distinguish between a law-side (order side) and a factual side. The sphere of spatial laws differs from the sphere of numerical laws – in an exemplary way expressed in the difference between 2+2=4 and 2+2=√8.

Remark: At this stage it should be mentioned that the aim of our analyses is not in the first place directed at interconnections between different mathematical sub-disciplines. The goal is to show that number and space are not only unique and irreducible aspects of (ontic) reality, but also to argue that they mutually cohere in many ways (eventually highlighted with reference to what will be designated as *analogies* on the law-side and on the factual side of these aspects). Whenever interconnections between mathematical sub-disciplines are highlighted the aim is to demonstrate the underlying ontic interconnections between the aspects of number and space.

7 Distance: highlighting the mutual coherence between number and space

We may now return to the mentioned key element in the modified definition given by Russell, *distance*: a line "is simply the *distance* between two points." The after-effect of the Greek geometrization of mathematics is seen in the long-standing and persistent use of the term 'Größe' ('magnitude' – for numbers) up to 19th century mathematicians – such as Bolzano and Cantor (in spite of their 'arithmetizing' intentions their designation of numbers still used the gateway of the spatial aspect). Greek mathematics already indirectly wrestled with *spatial magnitudes* – such as lengths, surfaces and volumes – although the *ratios* contemplated by them were treated in non-numerical contexts. By comparing spatial figures (such as line-segments, surfaces and solids) Euclidean geometry used ratios of magnitudes within a non-numerical context (sometimes a physical one) in their measurements. Naturally the Greeks were fully aware of specific *numerical properties* of spatial figures, because otherwise they would not have had a concept of a *triangle*, i.e. of a figure with *three* sides and *three* angles.

In itself this already shows that spatial figures (such as triangles) reveal an unbreakable coherence with the meaning of number. Of course it should be remembered that the overemphasis of number as a mode of explanation caused the Pythagoreans to see spatial figures *as numbers*. Kurt von Fritz remarks: "Likewise, so they said, 'are' the geometrical figures in reality the numbers or bundles of numbers that constitute the length relationships of their sides; through them their form is determined and through them they can therefore be expressed" (Von Fritz, 1965:287). It was only through the analytical geometry of Descartes and Fermat that numerical magnitudes were eventually contemplated – assigned to line segments, surfaces and solids. Savage & Ehrlich remarks: "Euclidean geometry compares lengths, areas, and regions by comparing physical, non-numerical *ratios* of these magnitudes and in effect uses such ratios in the place of our numbers" (see Savage & Ehrlich, 1992:1 ff.). However, the lacking understanding of the interconnections between number and space caused the mistaken identification of a line *with* its length (distance).

The first observation to be made in this connection is to establish that the notion of a 'line' as the 'distance' between two 'points' concerns *spatial* realities. A line is a spatial subject (configuration), not an arithmetical one. Yet the crucial question is: how can one designate the 'distance' between two points? The answer is: by specifying a *number* (for example by saying it is 3 inches long). The problem with this answer is that something *spatial*, namely a 'line', is now apparently equated with something *numerical*, namely 'distance'! In passing we note that the term 'distance' in yet a different way evinces an intrinsic connection with the meaning of number because a line is supposed to be the 'distance' between **two** points. Multiplicity ('two') is numerical; yet a multiplicity of *points* is **spatial**. Furthermore, the term 'inch' here has the function of the *unit of measurement*, i.e. the unit length. Therefore this unit is on a par with the notion of distance, because the number 1 and the number 3 respectively represent these two lengths. Does this mean that the domains of space and number are coinciding? If it is the case, then a question of priority arises (which embraces which one): is space numerical (then a 'line' is identical to 'distance', i.e. to number), or is number after all spatial in nature (then number, i.e. 'distance' is identical to space, i.e. to a 'line')? As we noted this concise dilemma reflects the basic contours of the history of mathematics as a discipline. After the initial Pythagorean claim that everything is number the discovery of irrational numbers turned mathematics into geometry. Then, during the 19th century arithmeticism once more gained the upper hand, although Frege close to the end of his life, reverted once more to the view that mathematics essentially is geometry. The situation is further complicated by the fact that the number specified (such as '3') does not stand on its own, i.e., it appears within a non-numerical context – one in which the general issue of *magnitude* prevails, with *length* as a one-dimensional magnitude. And to add insult to injury, we now suddenly have to account for another spatial notion: *dimensionality*! New problematic questions are now generated, for in our example of "3 inches" – related to the extension of a line – the reference to length brought with it the (spatial) perspective of *one* dimension (length specifies magnitude in the sense of one dimensional extension). On the one hand this points at *extension* which, presumably, essentially belongs to our awareness of space, while at the same time, just as in the case of the term 'distance', it reveals a connection with number as well, for one can speak about 1-dimensional extension (magnitude; i.e. of length), 2-dimensional extension (magnitude; i.e. of area), 3-dimensional extension (magnitude; i.e. of volume), and so on. Even if priority is given to the spatial context by admitting that the distinction between different dimensions is indeed something spatial, no one can deny that in some or other way number here plays a *foundational role*, for without number the given specification (regarding 1, 2, or 3 dimensions) is unthinkable.

Clearly, the term 'distance' is embedded within the domain of space and it also evinces a strict correlation between an *order of extension* (the law-side of this domain – i.e. *dimensionality*) and *factually extended spatial subjects* – spatial figures (such as 1-dimensional ones, i.e. lines), 2-dimensional ones, i.e. areas) and 3-dimensional ones (i.e. volumes).

The complexities generated by considering an order of extension correlated with factually extended spatial subjects (spatial figures) add weight to the suggestion that although something like a line has a spatial nature, its extension reveals that its true spatial meaning depends upon the meaning of number. The reason for this acknowledgement is found in the intrinsic role of numerical terms that are 'coloured' by space, such as *distance* and *dimension*. Within a numerical context, such as what is mathematically known as "real analysis," one can easily dispense with the concept of distance. But textbooks on real analysis sometimes still acknowledge that the geometric meaning of the term 'distance' may be useful, for "instead of saying that $|a-b|$ is 'small' we have the option of saying that a is 'near' b; instead of saying that '$|a-b|$ becomes arbitrarily small' we can say that 'a approaches b', etc." (Berberian, 1994:31)

8 Back to space

Two years after Russell gave his mentioned modified definition of a line as the *distance between two points*, the German mathematician, David Hilbert, published his axiomatic foundation of geometry: *Grundlagen der Geometrie* (1899). In this work Hilbert abstracts from the contents of his axioms, based upon three *undefined* terms: "point," "lies on," and "line." Suddenly the term 'distance' disappeared. The next year, when Hilbert attended the second international mathematical conference in Paris, he presented his famous 23 mathematical problems that co-directed the development of mathematics during the 20th century in a significant way – and in problem 4 he provides a formulation that opens up a new perspective on this issue, for instead of speaking of the *distance* between two points he talks of a straight line as the (shortest) *connection* of two points.[1] This choice of words completely avoids the traditional view, even found in the work of a contemporary mathematician like Mac Lane who still believes that the "straight line is the shortest distance between two points" (Mac Lane, 1986:17).

Hilbert's German term 'Verbindung' ('connection') does not define a line since it presupposes the meaning of continuous extension. Every part of a continuous line coheres with every adjacent part in the sense of being *connected* to it. Although it is tautological to say that the parts of a continuous line are fitted into a gapless coherence, it says nothing more than to affirm that the parts are *connected*. In this sense the connection of two distinct spatial points also highlights the presence of (continuous) spatial extension *between* the points that are connected to each other. In other words, Hilbert's formulation

---

[1] "[Das] Problem von der Geraden als kürzester Verbindung zweier Punkte" (see Hilbert, 1970: 302).

suggests that two points cannot be *connected* by a third point, but only by means of a line, i.e. through *continuous spatial extension*.

Combined with the primitive terms employed in his axiomatic foundation of geometry ('line', 'lies on' and 'point') the term 'connection' no longer equates a line with its distance. Once 'liberated' from this problematic bondage, alternative options emerge in order to account for the meaning of the term 'distance'. If *distance* is the 1-dimensional *measure* of factual (continuous spatial) extension, then one can do two things at once:
(i) acknowledge the *spatial* context of this measure (1-dimensional magnitude) and
(ii) account for the reference to number that is evident both in the '1' of 1-dimensional extension and in the (numerically specified) *length* evident in 'distance' as a specified (factual) spatial magnitude.

The core meaning of space, related to the awareness of extension and dimensionality, now acquires a new appreciation, further supported by the undefined nature of the term 'line' in Hilbert's 1899 work. The message is clear: if the core meaning of space (extension) is indefinable and primitive, then it is impossible to attempt to *define* a line by using a term revealing a reference to what is not *original* within space, namely the number (!) employed in the specification of the 'distance' between two points. On the one hand, distance as the *measure of extension* of a (straight) line depends upon and presupposes the existence of the line in its primitive 1-dimensional extension and can therefore never serve as a definition of it, and on the other hand it reveals a connection with the meaning of *number*. Therefore the 'definition' of a (straight) line as "the distance between two points" (Russell, Mac Lane) presupposes what it wants to define and consequently begs the question.

9 What is presupposed in space?

In our discussion of the question whether or not the domains of space and number are coinciding we have started by analyzing some consequences of the option that they do coincide. Aristotle already explored this possibility, but without success, because he employed the biological method of concept formation (of a *genus proximum* and *differentia specifica*) in a context where it does not fit. As *genus* his category of "quantity" is then differentiated into a *discrete quantity* and a *continuous quantity*: "Quantity is either discrete, or continuous" (*Categoriae*, 4 b 20). "Number, ... is a discrete quantity" (*Categoriae*, 4 b 31). The parts of a discrete quantity have no common limit, while it is possible in the case of a line (as a continuous quantity) to find a common limit to its parts time and again (*Categoriae*, 4b 25ff., 5 a 1ff.). In this account the aspects of number and space are brought under one umbrella and this approach precludes an insight into the uniqueness and irreducibility of number and space.

Rejecting Aristotle's approach calls for an acknowledgement of the fact that every specification of spatial configurations is unavoidably connected with terms reflecting in some or other way the coherence of space with the meaning of number (magnitudes and the number of dimensions). This outcome opens the way to the alternative option: of investigating the consequences of the assumption that although space and number are unique and distinct they still unbreakably cohere. The new question to be analyzed is then.

10 What is the interrelation between space and number?

If the measure of the factual (one dimensional) extension of a straight line could be specified by its distance, then the distance of a line not only presupposes its spatial extension since it also presupposes the intrinsic interconnection between the meaning of space and the meaning of number. Various mathematicians had an appreciation of this state of affairs.

But let us consider further options. The mere possibility to juxtapose two distinct 'facts', such as the statements that 2+2=4 and $\underline{2+2}=\sqrt{8}$, points in the direction of acknowledging two unique domains – each with its own sphere of laws and correlated subjects. Laugwitz refers to the approach of Bourbaki according to which there is a difference between what is discrete (algebraic structures) and what is continuous (toplogical structures).[1]

Of course modern mathematicians are inclined to give preference to the meaning of number and infinity. Tait remarks: "Surely the most important philosophical problem of Frege's time and ours, and one certainly connected with the investigation of the concept of number, is the clarification of the infinite, initiated by Bolzano and Cantor and seriously misunderstood by Frege" (Tait, 2005:213). What therefore needs to be clarified is summarized in the following two issues:

(i) which one of these two domains ('poles') is more fundamental, in the sense of *foundational*, to the other?

and

(ii) how should one account for the interconnections (interrelations) between these two domains ('poles')?

10.1 *Which region is more basic?*

Let us start with the approach of Bernays where he considers the way in which one can distinguish between our *arithmetical* and *geometrical* intuition. He rejects the widespread view that this distinction concerns *time* and *space*, for according to him the proper distinction needed is that between the *discrete* and the *continuous*.[2] Rucker also states: "The discrete and continuous repre-

---

1 .".. der Unterschied zwischen Diskretem (algebraische Strukturen) und Kontinuierlichem (topologische Strukturen)" (Laugwitz, 1986:12).

2 "Es empfiehlt sich, die Unterscheidung von 'arithmetischer' und 'geometrischer' Anschauung nicht nach den Momenten des Räumlichen und Zeitlichen, sondern im Hinblick auf den Unterschied des Diskreten und Kontinuierlichen vorzunehmen" (Bernays, 1976:81).

sent fundamentally different aspects of the mathematical universe" (Rucker, 1982:243). Fraenkel *et.al* even consider the relation between discreteness and continuity to be *the* central problem of the foundation of mathematics: "Bridging the gap between the domains of discreteness and of continuity, or between arithmetic and geometry, is a central, presumably even *the* central problem of the foundation of mathematics" (Fraenkel, A., et al., 1973:211). But then the question recurs: what is the relationship between the 'discrete' and 'continuous'?[1] In terms of the distinction between the domain of number and that of space the term "pattern" in the first place derives its meaning from *spatial configurations* or *patterns*. Only afterwards can one stretch this term – metaphorically or otherwise – in order to account for quantitative relations as well.

Whatever the case may be, speaking of a "discrete patterns" just as little bridges the gap between discreteness and continuity than referring to the "domain of number" does it (where the term "domain" is also derived from the meaning of space). The issue at stake in this connection is one falling outside the scope of this article for it concerns what should be treated in an analysis of the *elementary and compound basic concepts* of a scholarly discipline (such as mathematics).

Fraenkel *et.al*. even speak of a 'gap' in this regard and add that it has remained an "eternal spot of resistance and at the same time of overwhelming scientific importance in mathematics, philosophy, and even physics" (Fraenkel *et.al*., 1973:213). These authors furthermore point out that it is not obvious which one of these two regions – "so heterogeneous in their structures and in the appropriate methods of exploring" – should be taken as starting-point. Whereas the "discrete admits an easier access to logical analysis" (explaining according to them why "the tendency of arithmetization, already underlying Zenon's paradoxes may be perceived in [the] axiomatics of set theory"), the converse direction is also conceivable, "for intuition seems to comprehend the continuum at once," and "mainly for this reason Greek mathematics and philosophy were inclined to consider continuity to be the simpler concept" (Fraenkel *et.al*., 1973: 213).

Of course the modern tendency towards an arithmetized approach (particularly since the beginning of the 19th century) chose the alternative option by contemplating the primary role of number. Although Frege – as mentioned above – by the end of his life equated mathematics with geometry (consistent with the just mentioned position of Greek mathematics), his initial inclination certainly was to opt for the foundational position of number. Already in 1884

---

1 We noted above that the problem concerning which one is more basic – number or space – cannot be solved by specifying a *genus proximum* – albeit that of Aristotle with his distinction between a discrete quantity and a continuous quantity or that of the structuralist Resnik with his distinction between *discrete patterns* and *continuous patterns* (cf Aristotle: "Quantity is either discrete, or continuous" – *Categ*.4 b 20; and Resnik, 1997:201 ff. 224 ff.).

he asked if it is not the case that the basis of arithmetic is deeper than all our experiential knowledge and even deeper than that of geometry?[1]

From our discussion of the difference between an arithmetical and a spatial sum and in particular from our remarks about the term 'distance' it is possible to derive an alternative view on the order relation between the regions of discreteness and of continuity. Suppose we consider the idea that *discreteness* constitutes the core meaning of the domain of number and that *continuous extension* highlights the core meaning of space. Then these core meanings guarantee the distinctness or uniqueness of each domain. The domain of number, with its sphere of arithmetical laws and numerical subjects, is then seen as being stamped, characterized or qualified by this core meaning of *discreteness*. Likewise the domain of space, with its sphere of spatial laws and spatial subjects, is then viewed as being qualified by the core meaning of *continuous extension*.

But we have seen that a basic spatial subject, such as a (straight) line, cannot be understood without some or other reference to the meaning of number, for observing the *measure* of the line's extension requires the notion of 'distance' that involves number. Furthermore, since a line a spatial figure is extended in 1-dimension, it clearly only has a determinate meaning in subjection to the first order of spatial extension (namely *one* dimension). We have argued that in both domains (number and space) there is a strict correlation between the law-side and the factual side. In the case of space it is therefore possible to discern a reference to number both at the *law-side* and the *factual side*. Speaking of *one* or *more* than one dimensions presupposes the meaning of number on the law-side and this mode of speech at once specifies the meaning of the one dimensional extension, i.e. *magnitude*, of something like a line where the meaning of the number employed in the designation of the *length* of the line presupposes the original (primitive) meaning of number. The domain of number therefore appears to be more basic because an analysis of the meaning of space invariably calls upon foundational arithmetical features.

This conclusion is further supported by the approach of Maddy where she argues that most recent textbooks "view of set theory as a foundation of mathematics" (Maddy, 1997:22; see also Felgner, 1979:3) and that a set theoretic foundation can "isolate the mathematically relevant features of a mathematical object" in order to find a "set theoretic surrogate" for those features (Maddy, 1997:27, 34).[2] Bernays categorically asserts that "the representation of

---

1 "Liegt nicht der Grund der Arithmetik tiefer als der alles Erfahrungswissens, tiefer selbst als der der Geometrie?" (Frege, 1884:44).

2 Already in 1910 Grelling recognized set theory as the foundation of mathematics as a whole: "Zuerst ausgebildet als Hilfsmittel der Untersuchung bei gewissen Fragen der Analysis, hat sich die under den Händen inhres Schöpfers Georg Cantor und sein Schüler zu einer selbständigen metahmatischen Disziplin entwickelt, die heute die Grundlage der gesamten Mathematik bildet." ["In the first place developed as an auxilliary tool of the investigation of certain questions of analysis (set theory) in the hands of Cantor and his pupils (it was) developed into an independent mathematical discipline. Currently it constitutes the foundation of mathematics in its entirety" (Grelling, 1910:6).]

number is more elementary than geometrical representations" (Bernays, 1976:69).[1] In general one may view the arithmeticism of Weierstrass, Dedekind and Cantor as an (over-estimated) acknowledgement of the foundational position of the domain of number.

We may summarize the thrust of our preceding argument in favour of the foundational position of number in respect of space as follows:

The core meaning of space – namely *continuous extension* – entails *factual* extension in one or more *dimensions*; and specifying "one or more" dimensions presupposes the natural numbers 1, 2, 3, ... At the same time the 1-dimensional extension of a straight line comes to expression in the *measure* of this extension, designated as its *length* – and the latter (its *length*) is specified by using a *number* – showing that the meaning of spatial extension *intrinsically* presupposes ("builds upon") the meaning of number.

10.2 Interconnections between functional domains

A metaphorical way to capture this state of affairs is to use an image from human memory by saying that within the meaning of space (both at the law-side and the factual side), we discover configurations *reminding* us of the core meaning of number. A key element in all metaphorical descriptions is found in the connection between *similarities* and *differences*. Whenever what is different is shown in what is similar one may speak of *analogies*. Yet we want to broaden the scope of an analogy in order to include more than what is normally accounted for in a theory of metaphor. Our first designation already achieves this goal, for whenever *differences* between entities and properties bring to expression what is *similar* between those entities or properties, we meet instances of an *analogy*.[2] Implicit in the nature of an analogy is the distinction between something *original* and something else which 'reminds' one of what is originally given but that is now encountered in a *non-original* context, i.e. within an *analogical* setting. This is exactly what we have noticed in the terms 'distance' and 'dimension' – for in both cases the use of numerical terms in a spatial context *remind* us of their original (non-spatial) quantitative meaning. In terms of the idea of an analogy one can say that there is an analogy of number on the law-side of the spatial aspect (one, two, three or more dimensions) and that there is an analogy of number at the factual side of the spatial aspect (magnitude – as the correlate of different orders of extension: in *one* dimension magnitude appears as length, in *two* dimensions it appears as

---

1 He also states: "For our human understanding the concept of number is more immediate than the representation of space" (Bernays, 1976:75).

2 Whenever *entities* are involved in the figurative mode of speech such designations are considered to be *metaphorical*. But as soon as similarities and differences between modal functions (as they will be explained below) are captured, these purely aspectual interrelations represent a domain of analogies distinct from metaphors. When purely intermodal connections (analogies) are metaphorically explored, an element of the entitary dimension of reality will always be present (such as it is found in the metaphor of a person being 'reminded' of an original domain).

area, in *three* as volume). An account of the basic position of number can now be articulated in terms of the idea of analogies, for since basic *numerical analogies* are presupposed within the domain of space, the original meaning of number is indeed foundational to the meaning of space.

In the previous paragraph we introduced a new word in order to refer to the domains of number and space, namely the term 'aspect'. The underlying hypothesis of this usage is found in the theory that the various aspects of reality belong to a distinct dimension which is fundamentally different from the concrete *what-ness* of (natural and social) entities (such as things, plants, animals, artifacts, societal collectivities and human beings).

These concrete entities (and the processes in which they are involved) all function within the different aspects of reality. Questions about the way in which entities exist concern their *how-ness*, their mode of being. Aspects in this sense are therefore (ontic) *modes of being*. That my chair is *one* and has *four* legs reveal its function within the quantitative mode of reality; that it has a certain *shape* and *size* highlights its spatial function; that one can identify and distinguish it highlights its logical-analytical function, that it has a certain economic value demonstrates its function within the economic mode of reality, that it is beautiful or ugly brings to expression its aesthetic function, and so on.

This dimension of functions or aspects can also be designated as that of *modalities* or *modal functions*. What has already been said about the domains of number and space concern properties that may serve to define the nature of an aspect. Of course any description of modal aspects inevitably employs *metaphors* (involving entitary analogies). Fore example, one may say that aspects are 'points of entry' to reality, that they provide an 'angle of approach' to reality, and so on. Conversely, the modal aspects provide access to the dimension of entities – they may serve as *modes of explanation* of concrete reality.

Every aspect contains a sphere of modal (functional) laws (at its law-side); a factual side (subjected to modal laws); and a core meaning qualifying, characterizing or stamping all the structural moments discernable within an aspect (in particular also the analogical elements pointing to the meaning of other modal functions of reality). This core meaning or meaning-nucleus guarantees the uniqueness and irreducibility of every aspect and it underlies the inevitable use of primitive (= indefinable) terms by those disciplines that explore a specific modal aspect as angle of approach to reality. Some of these structural features of an aspect are captured in the sketch on the next page.

11 The irreducible meaning of space underlying Hilbert's primitive terms

Within the arithmetical aspect the factual relation between numbers is constituted as subject-subject relations – as it is present in the addition of numbers, the multiplication of numbers or establishing the numerical difference between numbers (subtraction). However, at the factual side of the spatial aspect there are not only subject-subject relations (such as intersecting lines), for there are also subject-object relations present, mainly expressed in the idea of a *boundary*.

Already in his abstraction theory Aristotle employed the notion of a boundary (or limit) – which is intuitively immediately associated with *spatial* notions (Aristotle used the term *eschaton*). By the 13th century AD Thomas Aquinas accounts for a 1-dimensional line by means of a descending series of abstractions. In contradistinction to natural bodies, all mathematical figures are infinitely divisible. The Aristotelian legacy is clearly seen in his definition of a point as the *principium* of a line (cf. *Summa Theologica*, I,II,2), which indicates the fact that a determinate line-stretch has points at its extremities ("cuius extremitates sunt duo puncta" – *Summa Theologica*, I,85,8). This legacy returns in a somewhat more general form in the 18th century (the era of the *Enlightenment*). Kant remarks:

> Area is the boundary of material space, although it is itself a space, a line is a space which is the boundary of an area, a point is the boundary of a line, although still a position in space (Kant, 1783, A:170).

In 1912 Poincaré discussed similar problems. Concerning the way in which geometers introduce the notion of three dimensions he says: "Usually they begin by defining surfaces as the boundaries of solids or pieces of space, lines as the boundaries of surfaces, points as the boundaries of lines" (cf. Hurewicz & Wallman, 1959:3). Although only related to three dimensions, Poincaré here provides us with an intuitive approach to dimension, implicitly stressing the unbreakable correlation between the law-side and the factual side in the spatial aspect:

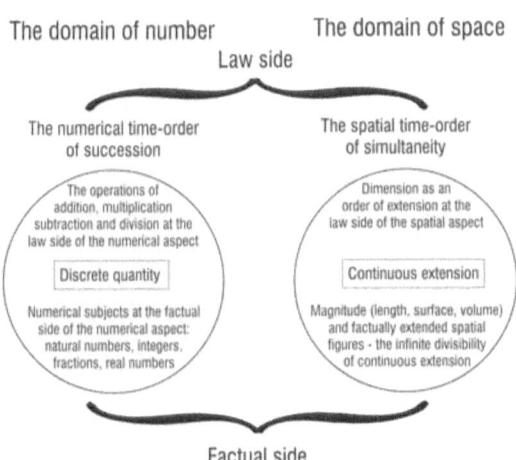

... if to divide a continuum it suffices to consider as cuts a certain number of elements all distinguishable from one another, we say that this continuum is of one dimension; if, on the contrary, to divide a continuum it is necessary to consider as cuts a system of elements themselves forming one or several continua, we shall say that this continuum is of several dimensions (Hurewicz & Wallman, 1959:3).

Before 1911 the problem of dimension was confronted with two astonishing discoveries. Cantor showed that the points of a line can be correlated one-to-one with the points of a plane, and Peano mapped an interval continuously on the whole of a square. The crucial question was whether, for example, the points of a plane could be mapped onto the points of an interval in both a continuous and one-to-one way. Such a mapping is called homeomorphic. The impossibility to establish a homeomorphic mapping between a "m-dimensional set and a (m+1)-dimensional set ($h > 0$)" was solved by Lüroth for the case where $m = 3$ (Brouwer, 1911:161). Brouwer provided the first general proof of the invariance of the number of a dimension (see Brouwer, 1911:161-165). Exploring suggestions of Poincaré, Brouwer introduced a precise (*topologically invariant*) definition of *dimension* in 1913, which was independently recreated and improved by Menger and Urysohn in 1922 (cf. Hurewicz & Wallman, 1959:4). Menger's formulation (still adopted by Hurewicz and Wallman) simply reads:

a) the empty set has dimension -1,
b) the dimension of a space is the least integer n for which every point has arbitrarily small neighborhoods whose boundaries have dimension less than n (Hurewicz & Wallman, 1959:4, cf. p.24).[1]

Whereas a spatial subject is always factually extended in some dimension (such as a 1-dimensional line, a 2-dimensional area, and so on), a spatial object merely serves as a *boundary* (in a delimiting way). The boundaries of a determined line-stretch are the two points delimiting it (with the line as a one-dimensional spatial subject). But these boundary points themselves are not extended in one dimension. Within one dimension points are therefore not spatial subjects but merely spatial objects, dependent upon the factual extension of the line. Yet a line may assume a similar delimiting role within two dimensions – for the lines delimiting an area are not themselves extended in a two dimensional sense. Likewise a surface can fulfil the role of a spatial object, namely when it delimits three dimensional spatial figures (such as a cube).

In general it can therefore be stated that whatever is a spatial subject in n dimensions is a spatial object in $n+1$ dimensions. A point is a spatial object in one dimension (an objective numerical analogy on the factual side of the spatial aspect), and therefore a spatial subject in *no* dimension (i.e. in *zero* dimensions). In terms of the fundamental difference between a spatial subject and a spatial object, it is impossible to deduce spatial extension from spatial objects (points). It is therefore unjustifiable to see a line as a set of points. But it falls outside the scope of this presentation to highlight the circularity present in

---

1   See also Brouwer, 1924:554. When a "species" π does not contain a continuum as part it is of dimension 0 in the Menger-Urysohn sense.

Grünbaum's attempt to argue for a consistent conception of the extended linear continuum as an aggregate of unextended elements (see Grünbaum, 1952). Grünbaum did not realize that the actual infinite – or, as we prefer to call it: the *at once infinite* – depends upon a crucial *spatial* feature, namely the *spatial order of simultaneity*. In the idea of the at *once infinite* the meaning of number points towards the meaning of space in an analogical way. Every known attempt to reduce space to number employs the at once infinite – and since the latter pre-supposes the irreducibility of the spatial order of at once these attempts all turn out to be *circular* (in the sense that one can reduce space to number if and only if one assumes the irreducibility of space). This remark also applies to the ideas advanced by Carl Posy in connection with building a "continuous manifold" out of "real numbers" (see Posy, 2005:321 ff.).

We can now account for the three primitive terms in Hilbert's axiomatization of geometry in the context of the spatial subject-object relation. The term 'line' reflects the primary existence of a (one dimensional) spatial *subject*, the term 'point' highlights the primary existence of a (one dimensional) spatial *object* and the phrase 'lies on' accounts for the *relation* between a spatial subject and a spatial object – in other words, it highlights the *spatial subject-object relation*.

Primitive features at the factual side of the spatial aspect	Subject	Object	Relation
The primtive terms in Hilbert's axiomatization of geometry (1899)	Line	Point	Lies on

From our discussion thus far it is clear that the theory of modal aspects constitutes a key element of a non-reductionist understanding of reality.

12 The theory of modal aspects

The theory of *modal law-spheres* first of all acknowledges the *ontic givenness* of the modal aspects. Hao Wang remarks that Gödel is very "fond of an observation that he attributes to Bernays": "That the flower has *five* petals is as much part of objective reality as that its color is *red*" (Wang, 1982:202). The quantitative side (aspect) of things (entities) is not a *product* of thought – at most human reflection can *explore* this given (functional) trait of reality by analyzing what is entailed in the *meaning* of multiplicity. Yet, in doing this (theoretical and non-theoretical) thought explores the *given* meaning of this quatitative aspect in various ways, normally first of all by *forming* (normally called: *creating*) **numerals** (i.e., number symbols). The simplest act of counting already explores the *ordinal meaning* of the quantitative aspect of reality. Frege correctly remarks "that counting itself rests on a one-one correlation, namely between the number-words from 1 to n and the objects of the set" (quoted by Dummett, 1995:144).

However, in the absence of a sound and thought-through distinction between the *dimension of concretely existing entities* (normally largely identified with 'physical' or "space-time existence") and the *dimension of functional modes* (aspects) of ontic reality, which cannot be observed through sensory perception, mathematicians oftentimes struggle to account for the epistemic status of their "subject matter." Perhaps the awareness for the need of acknowledging this distinct dimension of reality is best articulated in Wang's discussion of Gödel's thought. Wang discusses Gödel's ideas regarding "mathematical objects" and mentions his rejection of Kant's conception that they are 'subjective'. Gödel holds: "Rather they, too, may represent an aspect of objective reality, but, as opposed to the sensations, their presence in us may be due to another kind of relationship between ourselves and reality" (quoted by Wang, 1988:304, cf. p.205). To this Wang adds his support: "I am inclined to agree with Gödel, but do not know how to elaborate his assertions. I used to have trouble by the association of objective existence with having a fixed 'residence' in spacetime. But I now feel that 'an aspect of objective reality' can exist (and be 'perceived by semiperceptions') without its occupying a location in spacetime in the way physical objects do" (Wang, 1988:304).

Of course Wang could have referred to the important insights of Cassirer in this regard. Already in his article on Kant and modern mathematics (1907), and particularly in his influential work: *Substance and Function* (1910), Cassirer distinguishes between *entities* and *functions*. He clearly realizes that quantitative properties are not exhausted by any individual entity: ""number is to be called universal not because it is contained as a fixed property in every individual, but because it represents a constant condition of judgment concerning every individual as an individual" (Cassirer, 1953:34). If we set aside the (neo-)Kantian undertones of this statement, Cassirer already saw something of the *modal universality* of the arithmetical aspect of reality.

Every aspect has an undefinable core (or: nuclear) meaning (also designated as the meaning-nucleus) which *qualifies* all the analogical meaning-moments within a specific aspect. These analogical moments may refer backwards to *ontically earlier* aspects (known as *retrocipations*) or forwards to *ontically later* aspects (known as *anticipations*). *Earlier* and *later* are taken in the sense of the *cosmic time-order* as it is called by Dooyeweerd. The aspects of reality are fitted in an inter-modal coherence of earlier and later. The most basic aspect is that of number (meaning-nucleus: discrete quantity), which is followed by the aspect of space (continuous extension), the kinematical aspect (core: constancy), the physical (change/energy operation/interaction),[1]

---

[1] According to Sikkema the kernel of the physical aspect is "interaction" (Sikkema, 2005:20). "While some use force and/or energy to characterize the physical aspect, these concepts lose their meaning and relevanceat the quantum level, while interaction does not" (Sikkema, 2005:30). However, the expression *interaction* is compostie, and the first element, 'inter', represents a sptial analogy within the physical aspect and therefore cannot serve as a part of the characterization of the meaning-nucleus of this aspect. What is left is the second part, 'action', which is the equivalent of 'operation' in the expression *energy-operation*. If the term

the biotic (life), the sensitive (feeling), the logical (analysis), the cultural-historical (formative control/power), the sign-mode (symbolical signification), and so on. At the factual side of each aspect there are subject-object relations (except for the numerical aspect where within which there are only subject-subject relations).

The meaning of an aspect finds expression in its coherence with other aspects (retrocipations and anticipations). *Retrocipatory analogies* are captured in the *elementary basic concepts* of a discipline.

The first challenge in an analysis of the *elementary (analogical) basic concepts* of the various academic disciplines is to identify the modal "home" or "seat" of particular terms.

Within an aspect we discerned a difference between the *order-side* (also known as the law-side) and its correlate, the *factual side* (that which is subjected to the law-side and *delimited* and *determined* by the latter). The numerical time-order of succession belongs to the law-side of the arithmetical aspect, and any *ordered* sequence of numbers appears at its factual side (think of the natural numbers in their normal succession). With the exception of the numerical aspect (which only have subject-subject relations), all the other aspects in addition also have subject-object relations at their factual side.[1]

13 The impasse of arithmeticism

It is intuitively clear that our awareness of *succession* and *multiplicity* (underlying the concept of an ordinal number and induction) makes an appeal to the quantitative aspect of reality. These terms therefore have their modal "seat" ("home") in the arithmetical aspect.

Of course it is natural that special scientists will attempt to reduce apparently primitive terms to familiar and more basic ones. But if such an attempt becomes circular, or even worse, contradictory, then it may be the case that the primitive terms involved are truly *irreducible*! Phrased differently: an attempt to define what is undefinable may end up in *antinomic reduction*.[2] Sometimes the challenge is not to get *out* of the circle, but to get *into* it(s irreducible meaning)!

In the course of our preceding discussion the following cluster of terms probably transcend the confines of the numerical aspect: *simultaneity* (at once), *completedness*, *wholeness* (totality), and the *whole-parts relation*.

The most prominent recognition of the spatial "home" of wholeness and totality is found in the thought of Bernays. He writes that it is recommendable not to distinguish the arithmetical and geometrical intuition according to the

---

'energy' fails to apply to the quantum level, the expression energy-operation within such a context may be reduced just to 'operation' or 'action'. In a recent e-mail (May 24, 2008) Stafleu states his prefrence for *activity* (equivalent to the Greek *energeia*).

1 The identifiability and distinguishability of something represents its *latent* logical object-function. When it is identified and distinguished by a thinking subject, this analytical object-function is made *patent*.

2 The classical example is Zeno's attempt to define movement in static spatial terms.

moments of the spatial and the temporal, but rather by focusing on the difference between the *discrete* and the *continuous*.[1] Being fully aware of the arithmeticistic claims of modern analysis it is all the more significant that Bernays questions the attainability of this ideal of a *complete arithmetization* of mathematics. He categorically writes:

> We have to concede that the classical foundation of the theory of real numbers by Cantor and Dedekind does not constitute a *complete* arithmetization of mathematics. It is anyway very doubtful whether a complete arithmetization of the idea of the continuum could be fully justified. The idea of the continuum is after all originally a geometric idea (Bernays, 1976:187-188).[2]

Particularly in explaining the difference between the potential and the actual infinite the difference between *succession* and *at once* and the irreducibility of the notion of a *totality* surfaces. Hilbert introduces the difference between the potential and the actual (or: genuinely) infinite by using the example of the "totality of the numbers 1, 2, 3, 4, ..." which is viewed as a unity which is given at once (completed):

> If one wants to provide a brief characterization of the new conception of infinity introduced by Cantor, one can indeed say: in analysis where the infinitely small and the infinitely large feature as limit concept, as something becoming, originating and generated, that is, as it is stated, with the potential infinite. But this is not the true infinite. We have the latter when, for example, we view the totality of the numbers 1, 2, 3, 4, ... as a completed unity or when we observe the points of a line as a totality of things, given to us as completed. This kind of infinity is designated as the actual infinite.[3]

According to Lorenzen the understanding of real numbers with the aid of the actual infinite cannot camouflage its ties with space (geometry):

> The overwhelming appearance of the actual infinite in modern mathematics is therefore only understandable if one includes geometry in one's treatment. ... The actual infinite contained in the modern concept of real numbers still reveals its descent (Herkunft) from geometry (Lorenzen, 1968:97).

---

1 "Es empfiehlt sich, die Unterscheidung von "arithmetischer" und "geometrischer" Anschauung nicht nach den Momenten des Räumlichen und Zeitlichen, sondern im Hinblick auf den Unterschied des Diskreten und Kontinuierlichen vorzunehmen" (Bernays, 1976:81).

2 "Zuzugeben ist, daß die klassische Begründung der Theorie der reellen Zahlen durch Cantor und Dedekind keine *restlose* Arithmetisierung bildet. Jedoch, es ist sehr zweifelhaft, ob eine restlose Arithmetisierung der Idee des Kontinuums voll gerecht werden kann. Die Idee des Kontinuums ist, jedenfalls ursprünglich, eine geometrische Idee."

3 "Will man in Kürze die neue Auffassung des Unendlichen, der Cantor Eingang verschafft hat, charakterisieren, so könnte man wohl sagen: in der Analysis haben wir es nur mit dem Unendlichkleinen und dem Unendlichengroßen als Limesbegriff, als etwas Werdendem, Entstehendem, Erzeugtem, d.h., wie man sagt, mit dem potentiellen Unendlichen zu tun. Aber das eigentlich Unendliche selbst ist dies nicht. Dieses haben wir z. B., wenn wir die Gesamtheit der Zahlen 1, 2, 3, 4, ... selbst als eine fertige Einheit betrachten oder die Punkte einer Strecke als eine Gesamtheit von Dingen ansehen, die fertig vorliegt. Diese Art des Unendlichen wird als aktual unendlich bezeichnet" (Hilbert, 1925:167).

Lorenzen highlights the same assumption when he explains how real numbers are accounted for in terms of the actual infinite:

> One imagines much rather the real numbers as all at once actually present – even every real number is thus represented as an infinite decimal fraction, as if the infinitely many figures (Ziffern) existed all at once (alle auf einmal existierten) (Lorenzen, 1972:163).

Our discussion regarding $2+2=\sqrt{8}$ argued that within the quantitative aspect the order of succession (on its law-side) provides a basis for arithmetical operations such as addition and multiplication and their inverses and it also makes possible our basic numerical awareness of *greater* and *lesser*. The arithmetical order of succession therefore determines our most basic intuition of infinity, in the literal sense of one, another one, and so on, without an end, *endlessly, indefinitely, infinitely*. The traditional designation of this kind of infinity, known as the *potential infinite*, lacks an intuitive appeal. But when we alternatively refer to the 'successive infinite' this shortcoming is left behind. The other kind of infinity, traditionally known as the *actual infinite*, also calls for an "intuitively transparent" designation – such as the *at once infinite*. The successive infinite, presupposed in the infinite divisibility of continuity, makes possible induction, which, according to Weyl, guarantees that mathematics does not collapse into an enormous tautology (Weyl, 1966:86). According to Gödel non-"tautological" relations between mathematical concepts "appears above all in the circumstance that for the primitive terms of mathematics, axioms must be assumed" (Gödel, 1995:320-321). In the case of finitism where the "general concept of a set is *not* admitted in mathematics proper ... induction must be assumed as an axiom" (Gödel, 1995:321).

These modes of speech highlight the inevitability of employing terms with a *spatial descent* even when the pretention is to proceed purely in *numerical* terms. Lorenzen correctly points out that arithmetic by itself does not provide any motive for the introduction of the actual infinite (Lorenzen, 1972:159). The fundamental difference between arithmetic and analysis in its classical form, according to Körner, rests on the fact that the central concept of analysis, namely that of a real number, is defined with the aid of actual infinite totalities ("aktual unendlicher Gesamtheiten" – 1972:134). Without this supposition Cantor's proof on the non-denumerability of the real numbers collapses into denumerability. While rejecting the actual infinite, intuitionism interprets Cantor's diagonal proof of the non-denumerability of the real numbers in a constructive sense – cf. Heyting (1971:40), Fraenkel *et al.* (1973:256,272) and Fraenkel (1928:239 note 1). However, in order to reach the conclusion of non-denumerability, every constructive interpretation falls short – simply because there does not exist a *constructive* transition from the potential to the actual infinite (cf. Wolff, 1971).

It seems to be impossible to develop set theory without "borrowing" key-elements from our basic intuition of space, in particular the (order of) *at once* and its factual correlate: *wholeness / totality*. Since spatial subjects are *extended* their multiple parts exist all at once. This multiplicity is at the factual side of the spatial aspect a *retrocipation* to the meaning of number – i.e., *multiple parts* analogically reflect the meaning of number (multiplicity) within space.

Bernays did not have a theory of modal aspects at his disposal and therefore lacks the possibility of articulating explicitly the intermodal connections between number and space. For example, in stead of saying that the mathematical analysis of the *meaning of number* reveals an anticipation to the meaning of space, he states that the idea of the continuum is a geometrical idea which analysis expresses with an arithmetical language.[1]

When, under the guidance of our theoretical (i.e., modally abstracting) insight into the meaning of the spatial order of simultaneity, the original modal meaning of the numerical time-order is disclosed (deepened), we encounter the *regulatively deepened anticipatory* idea of actual or completed infinity. Any sequence of numbers may then, directed in an anticipatory way by the spatial order of simultaneity, be considered *as if* its infinite number of elements are present as a *whole* (*totality*) *all at once*.

In this context it is noteworthy that Hao Wang informs us that Kurt Gödel speaks of sets as being "quasi-spatial" and then adds that he is not sure whether Gödel would have said the "same thing of numbers" (1988:202). This mode of speech is in line with our suggestion that the undefined term "element of" employd in ZF set theory actually harbours the *totality* feature of continuity. The implication is that in an anticipatory way set theory is dependent on "something spatial"!

This also amounts to a confirmation of the unbreakable coherence between the law-side and the factual side of the numerical and the spatial aspects. The modal anticipation from the numerical time-order to the spatial time-order must therefore have its correlate at the factual side. At the factual side of the numerical aspect we first of all encounter the sequence of natural numbers (expressing the primitive meaning of numerical discreteness). Then there are the integers (keeping in mind that the term 'integer' derives from wholeness and therefore points forward to what is non-integral, namely fractions). Introducing the *dense* set of rational numbers imitates the infinite divisibility of spatial continuity. Since this divisibility embodies the successive infinite it represents, within space, a retrocipation to the numerical time-order of succession. Therefore, as an anticipation to a retrocipation,) the rational numbers represent the *semi-disclosed* meaning of number.

When we employ the anticipation at the law-side of the numerical aspect to the law-side of the spatial aspect we encounter the intermodal foundation of the notion of *actual infinity* – although the basic intuitions at play here are better served by the phrases suggested above, namely the successive and the at once ininfinite. The fact that the at once infinite *deepens* the meaning of number requires a brief remark explaining the "as if" character of this disclosed

---

1 "Die Idee des Kontinuums ist, jedenfalls ursprünglich, eine geometrische Idee, welche durch die Analysis in arithmetischer Sprache ausgedrükt wird" (Bernays, 1976:74).

notion of infinity. The anticipation from number to space on the law-side *determines* the multiplicity of natural numbers, integers and rational numbers which are correlated with it. Uor suggestion is that under the guidance of the actual infinite these sequences of numbers are considered *as if* they are present as completed (though infinite) *wholes* or *totalities* given at once.

Remark: "As if": the actual infinite as a regulative hypothesis

Vaihinger developed a whole philosophy of the "as if" (*Die Philosophie des Als Ob*), in which he tries to demonstrate that various special sciences may use, with a positive effect, certain *fictions* which in themselves are considered to be *internally antinomic*. The infinite, both in the sense of being infinitely large and infinitely small, is evaluated by Vaihinger as an example of a necessary and fruitful fiction (cf. Vaihinger, 1922:87 ff., p.530). Ludwig Fischer presents a more elaborate mathematical explanation of this notion of a *fiction*. In general he argues: "The definition of an irrational number by means of a formation rule always involves an 'endless', i.e. unfinished process. Supposing that the number is thus given, then one has to think of it as the completion (Vollendung) of this unfinished process. Only in this ... the internally antinomic (in sich widerspruchsvolle) and *fictitious* character of those numbers are already founded" (Fischer, 1933:113-114). Without the aid of a preceding analysis of the modal meaning of number and space, this conclusion is almost inevitable. Vaihinger and especially Fischer simply use the number concept of *uncompleted infinity* (the successive infinite) as a standard to judge the (onto-)logical status of the actual infinite. Surely, within the closed (not yet deepened) meaning of the numerical aspect, merely determined by the arithmetical time-order of *unfinished succession*, the notion of an actual infinite multiplicity indeed is *self-contradictory*.

However, the meaning intended by us for the actual infinite *transcends* the limits of this concept of number since, in a regulative way, it refers to the core meaning of the spatial aspect which (in an anticipatory sense) underlies the *hypothetical* use of the time-order of *simultaneity* (the "all" viewed as being present *at once*).

Paul Lorenzen echoes something of this approach in his remark that the meaning of actual infinity as attached to the "all" shows the employment of a fiction – "the fiction, as if infinitely many numbers are given" (Lorenzen, 1952:593). In this case too, we see that the "as if" is ruled out, or at least disqualified as something *fictitious*, with an implicit appeal to the primitive meaning of number.

As long as one sticks to the notion of a *process*, one is implicitly applying the yardstick of the *successive infinite* to judge the actual infinite.

Paul Bernays did see the essentially *hypothetical* character of the *opened up* meaning of number, without (due to the absence of an articulated analysis of the modal meaning coherence between number and space) being able to exploit it fully: "The position at which we have arrived in connection with the theory of the infinite may be seen as a kind of the philosophy of the 'as if'. Nevertheless, it distinguishes itself from the thus named philosophy of Vaihinger fundamentally by emphasizing the consistency and trustworthiness of this formation of ideas, where Vaihinger considered the demand for consistency as a prejudice ..." (Bernays, 1976:60).

Although the deepened meaning of infinity is sometimes designated by the phrase *completed infinity*, this habit may be misleading. If *succession* and *simultaneity* are irreducible, then the idea of an infinite totality cannot simply be seen as the completion of an *infinite succession*. When Dummett refers to the classical treatment of infinite structures "as if they could be completed and then surveyed in their totality" he equates this "infinite totality" with "the entire output of an infinite process" (1978:56). The idea of an infinite totality simply transcends the concept of the successive infinite.

A remarkable ambivalence in this regard is found in the thought of Abaraham Robinson. His exploration of *infinitesimals* is based upon the meaning of the at once infinite. A number a is called *infinitesimal* (or *infinitely small*) if its absolute value is less than m for all positive numbers m in $\Re$ ($\Re$ being the set of real numbers). According to this definition 0 is *infinitesimal*. The fact that the infinitesimal is merely the correlate of Cantor's transfinite numbers is apparent in that r (*not equal to* 0) is infinitesimal if and only if r *to the power minus 1* (r^{-1}) is infinite (cf. Robinson, 1966:55ff). In 1964 he holds that "infinite totalities do not exist in any sense of the word (i.e., either really or ideally). More precisely, any mention, or purported mention, of infinite totalities is, literally, *meaningless.*" Yet he believes that mathematics should proceed as usual, "i.e., we should act *as if* infinite totalities really existed" (Robinson, 1979:507).

Cantor explicitly describes the *actual infinite* as a constant quantity, *firm and determined in all its parts* (Cantor, 1962:401). Throughout the history of Western philosophy and mathematics, all supporters of the idea of *actual infinity* implicitly or explicitly employed *some* form of the *spatial order of simultaneity*. What should have been used as an *anticipatory regulative hypothesis* (the idea of *actual infinity*), was often (since Augustine) reserved for God or an eternal being, accredited with the ability to oversee any infinite multiplicity *all at once*.

This anticipatory regulative hypothesis of actual infinity does not *cancel* the original modal meaning of number, but only *deepens* it under the guidance of theoretical thought.

The new phrases for speaking of the *potential* and *actual infinite*, namely the successive and at once infinite, already had surfaced in the disputes of the early 14th century concerning the infinity of God.[1]

These new expressions relate directly to our basic numerical and spatial intuitions, viz., our awareness of *succession* and *simultaneity* – and their mutual irreducibility is based upon the irreducibility of the aspects of number and space.[2]

---

1   Compare the expressions *infinitum successivum* and *infinitum simultaneum* (Maier, 1964: 77-79).

2   Dooyeweerd did not accept the idea of the *at once infinite* (*actual infinity*) owing to the fact that he was strongly influenced by the intuitionistic mathematicians Brouwer and Weyl in this regard. Cf. Dooyeweerd, 1997-I:98-99 (footnote 1) and 1997-II:340 (footnote 1).

A truly deepened and disclosed account of the real numbers cannot be given without the aid of the *at once infinite*. That this *anticipatory coherence* between number and space always functioned prominently in a deepened account of the real numbers, may be shown from many sources. It will suffice to mention only *one* in this context. But before we do that we have to return briefly to the relationship between mathematics and logic.

14 The circularity entailed in set theoretical attempts to arithmetize continuity

The nuclear meaning of space is *indefinable*. If one tries to define the *indefinable* two equally objectionable options are open:

(i) either one ends up with a *tautology* – coherence, being connected, and so on, are all synonymous terms for *continuity* – or, even worse,
(ii) one becomes a victim of (antinomic) *reduction*, i.e. one tries to reduce what is indefinable to something familiar but distinct.

While the idea is ancient, modern Cantorian set theory again came up with the conviction that a spatial subject such as a particular line must simply be seen as an infinite (technically, a non-denumerably infinite) set of points.

If the points which constitute the one dimensional continuity of the line were themselves to possess any extension whatsoever, it would have had the absurd implication that the continuity of every point is again constituted by smaller points than the first type, although necessarily they also would have had some extension. This argument could be continued *ad infinitum*, implying that we would have to talk of points with an ever-diminishing "size." In reality such "diminishing" points do not at all refer to real points, since they simply reflect the nature of continuous extension, which as we have seen, is *infinitely divisible*. Such points build up space out of space.

Anything which has *factual extension* has a subject-function in the spatial aspect (such as a chair) or is a modal subject in space (such as a line, a surface, and so forth). A point in space, however, is always dependent on a spatial subject since it does not itself possess any extension (see our earlier discussion, pp. 19-21).

The length, surface or volume of a point is always zero – it has none of these. If the measure of one point is zero, then any number of points would still have a zero-measure. Even a(n denumerable) infinite set of points would never constitute any positive distance, since distance presupposes an extended subject.

Grünbaum has combined insights from the theory of point-sets (founded by Cantor) with general topological notions and with basic elements in modern dimension theory in order to arrive at an apparently consistent conception of the extended linear continuum as an aggregate of unextended elements (Grünbaum, 1952:288 ff.). From his analysis it is clear that he actually had "unextended unit point-sets" in mind and not simply a set of "unextended points" (Grünbaum, 1952: 295). Initially he starts with a non-metrical topological description and then, later on, introduces a suitable metric normally used for Euclidean spaces (point-sets). The all-important presupposition of this analysis is the acceptance of the linear Cantorean continuum (arranged in an order of magnitude, i.e. the class of all real numbers) (cf. Grünbaum, 1952:296).

On the basis of certain distance axioms, the real function $d(x,y)$ (called the distance of the points x,y which have the coordinates x_i, y_i) is used to define the length of a point-set constituting a finite interval on a straight line between two fixed points (the number of this distance is its *length*). For example, the length of a finite interval (a,b) is defined as the non-negative quantity $b - a$ (disregarding the question about the interval's being closed, open, or half-open). In the limiting case of $a = b$, the interval is called "degenerate" with length zero (in this case we have a set containing a single point) (cf. Grünbaum, 1952:296).

Furthermore, division as an operation is only defined on sets and not on their elements, implying that the *divisibility* of finite sets consists in the formation of proper non-empty *subsets* of these (surely, the degenerate interval is indivisible by virtue of its lack of a subset of the required kind) (Grünbaum, 1952:301). Finally, the following two propositions are asserted and are considered to the perfectly consistent:
"1. The line and intervals in it are infinitely divisible" and
"2. The line and intervals in it are each a union of indivisible degenerate intervals" (1952:301).

If we confront this analysis of Grünbaum with our characterization of the nature of the actual infinite (the at once infinite), we soon realize that his whole approach is *circular*. We have seen that, on the basis of the regulative hypothesis of the at once infinite, not only the set of real numbers but also the number of line segments having a common end point could be considered as *non-denumerable infinite totalities*. In the latter case (i.e., in the case of a group of line segments), we may identify, within the modal structure of space, a retrocipation to an anticipation (a mirror-image of the structure of the system of rational numbers). This retrocipation to an anticipation ultimately underlies Grünbaum's statement: "the Cantorean line can be said to be already actually infinitely divided" (Grünbaum, 1952:300).

Seemingly, the objection that any denumerable sum of degenerate intervals (with zero-length) must have a length of zero, does not invalidate Grünbaum's claim that a positive interval is the union of a continuum of degenerate intervals, because in the latter case we are confronted with a non-denumerable number of degenerate intervals – obviously, if we cannot enumerate them, we cannot *add* them to form their sum (for this reason, measure-theory also side-steps the mentioned objection, valid for the denumerable case). (Any attempted "addition" would leave out at least one of them.)

In this argumentatyion the irreducibility of the spatial time-order of simultaneity to the numerical time-order of succession is presupposed – ultimately dependent on the irreducibility of the modal meaning of space to that of number. From this it directly follows that the spatial whole-parts relation, determined by the spatial order of simultaneity, is also irreducible – explaining why the typical *totality* character of the continuum reveals an unavoidable circularity in the attempted purely arithmetical 'definition' of continuity. In other words, the modal meaning of space, qualified by the primitive meaning-nucleus of continuous extension (expressing itself at the law-side as a simultaneous order for extension and at the factual side as dimensionally determined extension – with or without a defined metric), not only implies that this meaning-nucleus of space is irreducible to number, but also that the spatial order of simultaneity at the law-side and the whole-parts relation at the factual side of the spatial aspect are ultimately irreducible. Therefore the attempt to reduce space to number is circular, for it has to employ the idea of the at once infinite which presupposes (in an anticipatory way) the irreducible meaning of space.

Although he did not pay attention to the law-side of the spatial aspect (obviously because he did not dispose of an articulated meaning-analysis of the structure of number and space), Paul Bernays does appreciate the irreducibility of the spatial whole-parts relation (the totality feature of spatial continuity) (Bernays, 1976:74).

> The property of being a totality "undeniably belongs to the geometric idea of the continuum. And it is this characteristic which resists a complete arithmetization of the continuum" ("Und es ist auch dieser Charakter, der einer vollkommenen Arithmetisierung des Kontinuums entgegensteht – 1976:74).

Laugwitz realized that the real numbers, in terms of Cantor's definition of a set, are still individually distinct and in this sense 'discrete'. According to him the set concept was designed in such a way that what is continuous withdraws itself from its grip, for according to Cantor a set concerns the uniting of well-distinguished entities, implying that the discrete still rules.[1] Although this objection actually shows that Laugwitz did not understand the difference between the successive and the at once infinite properly, in its own way it could be seen as an objection to the arithmeticistic claims of modern mathematics. In this regard Laugwitz implicitly supports Bernays's deeply felt reaction against the mistaken and one-sided nature of modern arithmeticism, expressed in his words:

> The arithmetizing monism in mathematics is an arbitrary thesis. The claim that the field of investigation of mathematics purely emerges from the representa-

---

[1] "Der Mengenbegriff ist von vornherein so angelegt worden, daß sich das Kontinuierliche seinem Zugriff entzieht, denn es soll sich nach Cantor bei einer Menge ja handeln um eine "Zusammenfassung wohlunterschiedener Dinge ... – das Diskrete herrscht" (Laugwitz, 1986:10). And on the next page we read: "So kommt man dazu, die Frage nach der Mächtigkeit der Menge der reellen Zahlen als 'Kontinuumproblem' zu bezeichnen. In dieser Auffassung wird der Unterschied zwischen Diskretem und Kontinuierlichem verwischt: Je zwei Teilpunkte sind wohl voneinander unterschieden, aber ihre Gesamtheit soll das Kontinuum repräsentieren; dieses würde also durch Diskretes dargestellt."

tion of number is not at all shown. Much rather, it is presumably the case that concepts such as a continuous curve and an area, and in particular the concepts used in topology, are not reducible to notions of number (Zahlvorstellungen) (Bernays, 1976:188).

The aritmeticistic claims of set theory are circular for proving the non-denumarability of the real numbers requires (as anticipatory hypothesis) the at once infinite which in turn presupposes the irreducibility of space. Therefore, space can be reduced to number if and only if it cannot be reduced to number.

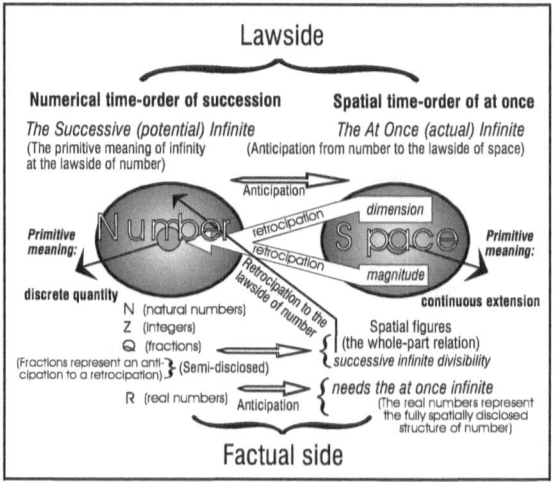

The mutual coherence and irreducibility of number and space

II
Physics

> It is sufficient to consider, for example, the discussions concerning the nature of space and time, determinism, indeterminism and causality, the continuous or discontinuous characteristics of matter and energy, the infinity or finiteness of the universe that have been produced by the developments of contemporary physics: even when they have been led by professional and outstanding physicists, they were of a genuine philosopical nature, as it can be easily seen if we consider that they also occupied the mind of several outstanding professional philosophers of our time. Moreover, they count among the most typical and classical questions of the philosophy of nature of all times, and the fact that they are still the object of lively discussions also after the acceptance, say, of relativity and quantum physics clearly indicates that they have not been *solved* by such theories, but simply further problematized as a consequence of them (Agazzi, 2001:10).

In addition to the aspects of number and space that played a central role in our discussion of the implications of a non-reductionist ontology an assessment of the foundation of the discipline of physics will have to analyze the meaning of the kinematic and physical aspects as well.

Interestingly the dominant philosophical orientation amongst the special sciences during the first half of the 20th century wanted to restrict science to the "positive facts" that were assumed to be the sole guide to "objective scientific truth." "Sense data," were supposed to be the only source of reliable knowledge, and it supported the postulate of the *neutrality* of human rational endeavors. The latter conviction (!) erroneously labelled any ultimate commitment (conviction) operative within the domain of rationality as a *disturbing factor* that should be *eliminated* from science.

However, without an implicit *trust* or *faith* in reason this postulate itself cannot be maintained. All human beings are endowed with the capacity to think and to argue rationally, but they do this from *diverging direction-giving orientations*. Consequently, in spite of acknowledging the universality of the structural conditions making possible human thinking in the first place, no single human being can escape from some or other deepest conviction. Stegmüller holds that there is no single domain in which a self-guarantee of human thinking exists – one already has to *believe* in something in order to *justify* something else (Stegmüller, 1969:314).

An analysis of the *structure* of scientific activities therefore does not aim at *securing* a domain of the *good* by protecting it from the evil influence of direction-giving ultimate commitments, for any such analysis can only proceed by implicitly proceeding from a particular life-orientation.

There are not simply 'scientific' people *liberated* from any and all supra-rational convictions, and "non-scientific" people blurred by the 'evil' of adhering to some or other *conviction*. Whatever the life-orientation of thinkers may be, they all equally share in the dimension of *rationality* (or: logicality) and all of them are inevitably in the *grip* of a more-than-rational *ultimate commitment*.[1]

The positivistic appeal to *sense data* is problematic, because the theoretical 'tools' employed in the *description* of what is observed always utilize *terms* that are not susceptible to "empirical observation" themselves. In order to demonstrate this point it will be instructive to consider the history of the concept of *matter*.

15 Historical perspective on the concept of matter

The early Greek philosophers have chosen some or other *fluid* element as principle of origin, such as *water* (Thales), *fire* (Heraclitus) and *air* (Anaximenes). Of course the subsequent development should take into account the significant role of the school of Pythagoras. The contribution of this school is that it articulated the insight that *rational knowledge* cannot be divorced from *numerical relationships*. Naturally this school went too far in its above-mentioned one-sided claim that *everything is number*. This thesis rests on the conviction that with the aid of the relation between integers, i.e. by merely using normal fractions, it is possible to describe the 'essence' of whatever there is in numerical terms.

However, soon the developments within Greek culture became sensitive to *spatial configurations* – such as the shape of the calyx leafs found in nature, for this shape appeared as an instantiation of a *regular pentagram*. An investigation of the geometrical properties of a regular pentagram led to the discovery that it is not possible to express the *ratio* between any side and any diagonal of the regular pentagram with the aid of normal fractions, i.e. in terms of the *ratio* of two whole numbers / integers: a/b. This limitation at once embodied the discovery of 'incommensurable' quantities – something completely unacceptable for the Pythagoreans because suddenly within the limiting and form-giving nature of number the *apeiron* (the unbounded-infinite) appeared, i.e. irrational numbers were discovered.

Flowing from the inherent tension in Greek thought between what *is limited* and what *is unlimited* (the *peras* and the *apeiron*) the discovery of irrational numbers (or in modern mathematical terms: real numbers) inspired the search for an alternative principle of explanation – one that can escape from the unbounded (infinite) present in number.

---

[1] A penetrating analysis is given by Clouser in terms of the hidden role of religious belief in theories (see Clouser, 2005).

The alternative mode of explanation that entered the scene was found in *space*. The spatial aspect allowed for the acceptance of *static forms* and it also opened the possibility to observe any spatial figure *at once*, without any *before* and *after*. The implication was that the acquisition of concepts is enclosed within the *now* and in the school of Parmenides this resulted, as we have seen, in the equation of *thought* and *being*.

It is known that on the basis of Babylonian observations Thales accurately predicted an eclipse of the sun in the year 585 B.C. He also had the remarkable geometrical skill to calculate the height of a pyramid from a sun shadow of 45^0 (keeping in mind that a pyramid differs from something like a tree where it is possible to establish its height perpendicular to its base). Thales also knew that the diagonals of a rectangular triangle are equidistant and according to Lorenzen he provides the starting-point for geometry as a coherent theoretical system (Lorenzen, 1960:45-46).

The important feature of this development is that the spatial figures of Greek geometry were *idealized*. It meant that a straight line, circle and square are not perceivable in a sensory way – they can merely be contemplated intellectually. Plato's account of human knowledge reflects this conviction because he explicitly states that the conclusions reached do not use "sensory objects":

> Then by the second section of the intelligible world you may understand me to mean all that unaided reasoning apprehends by the power of dialectic, when it treats its assumptions, not as first principles, but as *hypotheses* in the literal sense, things 'laid down' like a flight of steps up which it may mount all the way to something that is not hypothetical, the first principle of all; and having grasped this, may turn back and, holding on to the consequences which depend upon it, descend at last to a conclusion, never making use of any sensible object, but only of Forms, moving through Forms from one to another, and ending with Forms (*Politeia*, 510D).

Plato's dialogue Meno, where the leader of the conversation used leading questions in order to allow the conversation partner to produce a geometrical proof, caused Oskar Becker to remark that this gave birth to the appreciation of the *a priori* nature of mathematics (Becker, 1965:X).

The effect of the discovery of irrational numbers was not only that mathematics was geometrized for it also paved the way for a speculative theory of reality attempting to explain the entire universe in terms of a *spatial perspective* – as a substitute for the outdated arithmetical orientation of the Pythagoreans. The implication was that Greek thought now understood *matter* in terms of spatial extension. An entity is identified with the place it occupies. Something *is* its place.

It should be noted, however, that Parmenides hardly disposed over an independent space concept. He also did not contemplate the idea of an *empty space*. When something *is* its place, then the absence of something implies that the subject to which the predicate 'place' applies is not present. Herman

Fränkel writes: "With the assertion of a complete filling of space ... the existence of a mere empty space is rather denied than acknowledged." (Fränkel, 1968:181, note 4).

In its denial of movement the school of Parmenides, in particular the arguments of Zeno, merely formulated the consequences of over-emphasizing the spatial aspect as mode of explanation. If something indeed *is* its place then it can never move, for passing from one place to another place will entail a change of essence!

The metaphysical overextension of the *static* nature of space even motivated a remarkable denial of the spatial whole-parts relation.

In order to understand this properly we have to keep in mind what we have explained earlier, namely that whatever is continuously extended in a spatial sense allows for an infinite divisibility (see page 25 above). The spatial whole-parts relation turns the original numerical meaning of succession – the successive infinite – 'inwards', embodied in the successive infinite divisibility of a continuum. In terms of the inter-modal coherence between number and space and in the light of the foundational role of number it indeed belongs to the meaning of the spatial whole-parts relation that it contains the possibility of endless divisions.

The spatial metaphysics of Parmenides, for that matter, inspired Zeno to defend a view of *unitary wholeness* that *excludes* plurality. In other words, Zeno wants to *deny* the 'part'-element of the spatial whole-parts relationship while at the same time holding on to the 'wholeness' which entails it.

> His position is that reality is both *one* and *indivisible*. Yet, in order to *argue* for his position, he explored the whole-parts relation in his argument that is aimed at the *denial* of plurality! The reason why Zeno considers plurality to be self-contradictory is that plurality requires a *number* of (indivisible) units and because it also implies that reality is *divisible* (see Guthrie, 1980:88). But divisibility threatens the wholeness of a *unit*, since anything divisible has to be a magnitude which must be infinitely divisible. The supposed *indivisibility* of a *unit* clashes with its *infinite divisibility*. "Hence, since plurality is a plurality of units, there can be no plurality either" (Guthrie, 1980:89).

The antinomies of Zeno (including those of Achilles and the tortoise and the flying arrow) represent the strating-point of a long speculative tradition in which the meaning of space was metaphysically explored within the context of a speculative theory of *being* that finds in God – as the Highest Being (*ipsum esse*) – its conclusion.

By exchanging two *modes of explanation* and attempt to strip them from their intrinsic connections caused multiple distortions. The school of Parmenides did realize that space provides an original mode of explanation but in the attempt to 'purify' space from number it challenged a foundational condition of space, given in the nature and meaning of a *multiplicity*. By ignoring the foundational role of a numerical multiplicity Zeno distorted the meaning of number and at once also skewed the meaning of space by questioning the divisibility of a spatial continuum. The infinite divisibility of a continuous whole within space is a reminder of the original successive meaning of number lying at the basis of space. Just as little as it is possible to separate space from other aspects of reality can it be separated from the numerical aspect. Even in the most extreme examples of arithmeticism in modern mathematics, aiming at reducing space to number, key features of space were needed. In the case of axiomatic set theory it turned out to be unavoidable to use undefined ('primitive') terms derived from the spatial whole-parts relation, such as *set* or *element of* (see Fraenkel *et.al.*, 1973:21 ff).

The original numerical meaning of the number one as an integer analogically appears within the spatial aspect. The *unity* of a spatial subject is found in its *wholeness*. In other words, a spatial unity is constituted as a genuine *whole* or *totality*, a unitary whole allowing an *infinite divisibility*. The speculative (metaphysical) idea of a unitary whole *precluding* multiplicity robs both number and space from their unique meaning as well as from their mutual coherence.

In respect of the nature of material things the most important consequence is that the Greek-Medieval legacy only acknowledges *concrete material extension*. Extension characterizes the nature of material things.

Within the Aristotelian legacy it was believed that celestial bodies obey laws that are different from those that hold for entities on earth. In addition it was believed that the movement of anything required a *cause*. The problem of motion increasingly acquired a more prominent position, although it did not mean that the powerful influence of the classical space metaphysics immediately lost its hold. The power of this spatial orientation is indeed still evident in the thought of Descartes (1596-1650) and even Immanuel Kant (1724-1804). In their understanding of nature both these philosophers continued to assign a decisive role to *spatial extension*. For Descartes *extension* serves as the essential characteristic of material bodies – *res extensa*, for he writes: "That the nature of body consists not in weight, hardness, colour, and the like, but in extension alone" (Descartes, 1965:200 – Part I, IV). Kant's characterization of material bodies is also oriented to space. When our understanding leaves aside everything accompanying their representation, such as substance, force, divisibility, etc., and likewise also separate that which belong to sensation, such as impenetrability, hardness, color, etc., then from this empirical intuition something else is left, namely extension and shape.[1]

It should not surprise us therefore that Descartes straightaway applied the feature of (mathematical) continuity to material things and even to atoms that since Greek antiquity were supposed to be the last indivisible material parti-

---

1 "So, wenn ich von der Vorstellung eines Körpers das, was der Verstand davon denkt, als Substanz, Kraft, Teil-barkeit usw., imgleichen, was davon zur Empfindung gehört, als Undurchdringlichkeit, Härte, Farbe usw. absondere, so bleibt mir aus dieser empirischen Anschauung noch etwas übrig, nämlich Ausdehnung und Gestalt" (Kant, 1781/1787-B:35).

cles. He holds that there cannot be atoms or material particles that are inherently non-divisible.

> We likewise discover that there cannot exist any atoms or parts of matter that are of their own nature indivisible (Descartes, 1965:209; Part I, XX).

In this context (XX) he even introduces the idea of God in order to make acceptable the infinite divisibility of matter. He argues that although God can make a particle small enough that no creature can divide it this does not set any limits to the Divine capacity to divide. Therefore it should be assumed that matter is indeed infinitely divisible:

> Wherefore, absolutely speaking, the smallest extended particle is always divisible, since it is such of its very nature.

That Descartes continues to hold on to *extension* as the essential trait of matter embodies his connection with the long-standing Greek-Medieval tradition. However, what he has to say regarding the infinite divisibility of matter breathes the spirit of the early modern *functionalistic* orientation.[1] This new functionalistic attitude soon attempted to explain concrete things completely in *functional* terms. Yet Descartes at the same time did pay attention to motion, which he defined as the "action by which a body passes from one place to another" (Descartes, 1965:210; Part I, XXIV). This new point of view finds itself on the cross-road of the transition from the Medieval to the modern era.

Although Buridan (early 14th century) did contribute to the uprooting of the dominant position of spatial extension and the transition to the modern era, it should be kept in mind that the *impetus* idea itself cannot be equated with the nature of *inertia*. There is a conception that the mechanics of Buridan and classical physics are fundamentally similar, entailing that the *impetus theory* practically already brought to expression the *law of inertia*.

This convergence is first of all sought in the supposed correlation between the scholastic view of impetus and the dynamic element of inertial motion. In the second place it was believed that the assumption of *permanence* present in Buridan's view of impetus already discovered the law according to which motion not disturbed in any way will be everlasting (cf. Maier, 1949:142).

It is indeed striking that an *impetus* that was transferred in a celestial sphere (supposed to follow a *circular* path) was supposed to be free from any resistance. Therefore it seemed proper to compare it with the underlying idea of inertia. But inertia concerns uniform (rectilinear) motion, not circular motion. In the case of the impetus theory there was a difference between what happened in the heavens and happened on earth. According to the Medieval Scholastic understanding a force without resistance cannot produce motion. But since *impetus* is artificially and forcefully superimposed upon some or other obstacle (i.e. an interfering force of motion), such an obstacle can only be overcome when the impetus itself is altered in the process of overcoming the resistance. Yet the decisive difference between these two views is given in that element

---
[1] Functionalism reduces entities to functions and substantialism reduces functions to entities.

of inertial movement from which one cannot abstract, namely the inertia of a mass-point.[1] It is possible to abstract from external obstacles and forces, but it is impossible to abstract from that which is crucial for inertia, namely the *mass* of whatever moves. According the classical mechanics the latter is the real factor in the continuation of movement (inertia). In the case of the impetus theory inertial mass serves as resistance (obstacle) for the movement of the impetus that caused it. Consequently there is an unbridgeable gap between the impetus theory and the basic idea if inertia, namely the possibility of an everlasting rectilinear motion.[2]

Galileo formulated his law of inertia with the aid of a *thought experiment*. Suppose a body moves on a friction-free path extended into infinity, then this movement will simply continue endlessly. Opposed to the traditional Aristotelian-Scholastic conception according to which the movement of a body is dependent upon a *causing force* the law of inertia implies the motion is something *given* and that therefore in stead of trying to deduce or explain it it should be accepted as a mode of explanation in its own right. Motion is original and unique and indeed embodies a distinct *mode of explanation* different from those used by the Pythagoreans (number) and the Eleatic school of Parmenides (space). If motion does not need a causing force, then at most it is possible to speak of a *change* of motion (*acceleration* or *deceleration*) – and this does need a *physical force*. The well-known German physicist remarks:

> Since the law of inertia has shown that no force is required for a change of place the most natural thing to do is to accept that force causes a change of speed, or, as Newton says, the magnitude of motion ('Bewegungsgröße') (Von Weizsäcker, 2002:172).[3]

The idea of a uniform (rectilinear) motion on the one hand expands the inherent limitations attached to number and space as modes of explanation, and on the other it at once opens the way to consider another problem that already captured Greek thought. This problem concerns the relation between *persistence* (think about the nature of inertia) and *dynamics* (consider the change of motion requiring a physical force).

The important insight of Plato is that change can only be established on the basis of constancy (persistence) – i.e. without an enduring subject there is nothing to "hold on to," nothing to which the alleged changes can be attributed. Of course this insight does not force us to join the speculative account which Plato gave for it in his metaphysical theory of static, super-sensory ideal forms – although it is true that his solution did form a lasting attraction

---

1 In classical mechanics the simplest subject is a mass-point.
2 Maier writes: "Es gibt also gar keinen Ausweg: die Möglichkeit einer in infinitum daurenden gleichförmigen Bewegung des proiectum ist von Standpunkt der Impetustheorie aus grundsätzlich ausgeschlossen" (Maier, 1949:148).
3 "Da das Trägheitsgesetz gezeigt hat, daß keine Kraft nötig ist für eine Änderung des Orts, ist es das natürlichste, anzunehmen, die Kraft verursacht eine Änderung der Geschwindigkeit, oder, wie Newton sagt, der Bewegungsgröße."

for many scholars. Even Frege said that amidst the on-going flow of events something lasting, something with eternal durability must exist for otherwise the knowability of the world would be canceled and everything would collapse in confusion.[1]

The proper elaboration of Plato's insight, namely that change presupposes constancy, is found in Galileo's formulation of the law of inertia and in Einstien's theory of relativity. The core idea of Einstein's theory is after all die constancy of the velocity of light in a vacuum. Although he often merely speaks of "the principle of the constancy of the speed of light",[2] he naturally intends "the principle of the vacuum-velocity" ("das Prinzip der Vakuumlichtgeschwindigkeit" – see Einstein, 1982:30-31; and also Einstein, 1959:54). From this it follows that Einstein primarily aimed at a theory of *constancy* – whatever moves move relative to this element of constancy. It was merely a concession to the historicistic *Zeitgeist* at the beginning of the 20th century that he gave prominence to the term 'relativity' – all movement is relative to the constant c.

However, a certain ambiguity is still found in the thought of Descartes and his followers for in spite of the fact that they viewed *extension* as the essential property of matter, they also simultaneously pursued the kinematical ideal to explain everything that exists and happens exclusively in terms of movement (cf. Maier, 1949:143).[3] It is generally known that Thomas Hobbes took the full step to the exploration of movement as principle of explanation in his intended rational reconstruction of reality. According to the newly established natural science ideal he first demolished reality to a heap of chaos in order afterwards to build up, step by step, a new rationally ordered cosmos, guided by the key concept "moving body." His acquaintance with the mechanics of Galileo enabled him to exceed the limits of space as mode of explanation. Galileo himself embodies the long history of our understanding of matter up to this phase of its development because he explicitly explores the three modes of explanation thus far highlighted in our discussion. He accounts for arithmetical properties (countability), geometrical properties (form, size, position and contact) and kinematic features (motion).[4] Leibniz continues this legacy in his be-

---

[1] "Wenn in dem beständigen Flusse aller Dinge nichts Festes, Ewiges beharrte, würde die Erkennbarkeit der Welt aufhören und alles in Verwirrung stürtzen" (Frege, 1884:VII – Einleitung).

[2] "das Prinzip der Konstanz der Lichgeschwindigkeit" – cf. Einstein, 1982:32.

[3] Maier remarks that "Descartes und seine Schule" indeed pursued a "rein phoronomisches Ideal" and attempted to explain "alles Sein und Geschehen in der Welt lediglich aus Bewegungen."

[4] "G. Galilei zählt als primäre Qualitäten der Materie arithmetische (Zählbarkeit), geometrische (Gestalt, Größe, Lage, Berührung) und kinematische Eigenschaften (Beweglichkeit) auf" (Hucklenbroich, 1980:291).

lief that physical events can be explained mechanisticlly in terms of magnitude, figure, and motion.[1]

Writing on the foundations of physics, David Hilbert refers to the mechanistic ideal of unity in physics but immediately adds the remark that we now finally have to free ourselves from this untenable ideal (cf. Hilbert, 1970: 258).[2]

As soon as the kinematic mode of explanation is acknowledged in its own right the necessity to find a cause for motion disappears. The classical opposition between *being at rest* and *moving* is therefore untenable, because from a kinematic perspective 'rest' is a state of movement (cf. Stafleu, 1987:58). Unique and irreducible modes of explanation are not *opposites* – for they are mutually cohering and irreducible.[3]

Although Descartes and Newton did employ the concept *force*, it may in general be said that modern physics since Newton is characterized by its *mechanistic* main tendency. The mechanistic view consistently attempts to reduce all physical phenomena to a *kinematic perspective*. However, already in the course of the 19th century modern physics started to explore the nature of *energy*. The founder of physical chemistry, Wilhelm Ostwald, developed his so-called *Energetik* (enegetics) that even influenced the later views of Heisenberg. Vogel refers to Heisenberg's work "Wandlungen in den Grundlagen der Naturwissenschaft" (Stuttgart 1949) where the latter explicitly speaks of energy as the basic stuff that constitutes matter in its threefold stable forms: electrons, protons and neutrons (Vogel, 1961:37). Yet Ostwald's *Energetik* did not exert a lasting influence upon the physics of the 20th century, probably because it was attached to a specific view of continuity opposed to an atomistic approach. Niels Bohr particularly mentions the excessive skepsis found in the thought of Mach regarding the existence of atoms.[4]

The last prominent physicist who consistently adhered to the mechanistic approach was Heinrich Herz. Soon after Hertz's death in 1894 the work in which he attempted to restrict the discipline of physics to the concepts mass, space and time, reflecting the three most basic modes of explanation of reality,

---

1 On October 9, 1687 Leibniz wrote in a letter that we "must always explain nature mathematically and mechanically" (Leibniz, 1976:38). In a footnote the Editor of Leibniz's work writes that Leibniz's approval of the corpuscular philosophy of Boyle ought to be understood as "any philosophy which explains physical events mechanistically or in terms of magnitude, figure, and motion" (Leibniz, 1976:349, note 14).

2 It is therefore strange that the contemporary physical scientist from Cambridge, Stephen Hawking, still writes: "The eventual goal of science is to provide a single theory that describes the whole universe" (Hawking, 1988:10).

3 For that reason we have noted that also number and space ought not to be seen as opposites as it was asserted by Lakoff en Núñez (2000:324) owing to their inability to appreciate the unique and mutually cohering nature of these aspects.

4 See Niels Bohr, Atomtheorie und Naturbeschreibung (Berlin 1931:60 and p. 12), quoted by Vogel, 1961:35). It should be kept in mind that the views of Mach ought to be understood against the background of the position of Ostwald.

namely the modes of number, space and movement, appeared: "The Principles of Mechanics developed in a New Context." This caused him (and Russell) to view the concept of *force* as something intrinsically antinomous.

The Latin designation of *mass* during the medieval period was "quantitas materiae" (see Maier, 1949:144). From this it appears that number (quantitas) plays a key role in the concept *mass*. Mass concerns a *physical* quantity, but it is also possible to observe the quantity of energy from the perspective of the *kinematical* modality. In this case the technical expression is *kinetic energy* that indicates the action capacity inherent to a moving body (see Maier, 1949:142).

As soon as the physical aspect of reality surfaced it opened up the way for 20th century physics to explore it as an independent mode of explanation and to arrive at an even more nuanced understanding of reality. For example, in his *protohpysics* Paul Lorenzen distinguishes four units of measurement reflecting the first four modes of explanation: *mass*, *length*, *duration* and *charge* (Lorenzen, 1976:1 ff.).

A decade after Max Planck discovered his "Wirkungsquantum" he explicitly addressed the intrinsic untenability of the mechanical understanding of reality.

> The conception of nature that rendered the most significant service to physics up till the present is undoubtedly the mechanical. If we consider that this standpoint proceeds from the assumption that all qualitative differences are ultimately explicable by motions, then we may well define the mechanistic conception as the conviction that all physical processes could be *reduced completely to the motions* (the italics are mine – DFMS) of unchangeable, similar mass-points or mass-elements.[1]

Einstein is equally explicit in his negative attitude towards "the mechanistic framework of classical physics" (see Einstein, 1985:146).

Eventually the distinction between the kinematic and physical aspects of reality thus became common knowledge. According to Janich the scope of an exact distinction between *phoronomic* (subsequently called *kinematic*) and *dynamic* arguments could be explained in terms of an example. Modern physics has to employ a dynamic interpretation of the statement that a body can alter its speed only continuously. Given certain conditions a body can never accelerate in a discontinuous way, that is to say, it cannot change its speed

---

1 "Diejenige Naturanschauung, die bisher der Physik die wichtigsten Dienste geleistet hat, ist unstreitig die mechanische. Bedenken wir, daß dieselbe darauf ausgeht, alle qualitativen Unterschiede in letzter Linie zu erklären durch Bewegungen, so dürfen wir die mechanische Naturanschauung wohl definieren als die Ansicht, daß alle physikalischen Vorgänge sich vollständig auf Bewegungen von unveränderlichen, gleichartigen Massenpunkten oder Massenelementen zurückführen lassen" (Planck, 1973:53).

through an infinitely large acceleration, because that will require an infinite force.[1]

The idea of an attracting force, initially conceived of in connection with magnetism, eventually brought Newton to the insight that magnetism is a force that cannot be explained through motion, although in its own right, foundational to the physical aspect, motion is a mode of explanation. Stafleu points out that the rejection of the Aristotelian distinction between the physics of celestial bodies and the physics of things on earth paved the way, in the footsteps of Galileo and Descartes, to realize that the same physical laws apply to both domains, i.e. that physical laws display modal universality (i.e. they hold universally) (Stafleu, 1987:73). He also remarks that Newton (just as Kepler) indeed already appreciated *force* positively as a principle of explanation that is distinct from motion as an original principle of explanation (see Stafleu, 1987:76). Stafleu summarizes this process through which the physical aspect emerged as an equally original mode of explanation as follows:

> In Newtonian mechanics, a force is considered a relation between two bodies, irreducible to other relations like quantity of matter, spatial distance, or relative motion. Though an actual force may partly depend on mass or spatial distance, as is the case with gravitational force, or on relative motion, as is the case with friction, a force is conceptually different from numerical, spatial or kinematic relations (Stafleu, 1987:79).

Since the introduction of the atom theory of Niels Bohr in 1913, and actually already since the discovery of radio-activity in 1896 and the discovery of the energy quantum h, modern physics realized that matter is indeed characterized by physical energy operation. It is therefore understandable that 20th century physics eventually had to come to a general acknowledgement of the decisive significance of *energy operation* for the nature and understanding of the physical world, as it is strikingly captured in Einstein's famous formula:

$$E = mc^2$$

It was also realized that physical processes are *irreversible*. In itself this observation also justifies the distinction between the kinematic and the physical aspects of reality. Both Planck and Einstein knew that in terms of a purely kinematic perspective all processes are *reversible*. Einstein refers to Boltzmann

---

[1] "Die Tragweite einer strengen Unterscheidung phoronomischer (im folgenden kinematisch genannt) und dynamischer Argumente möchte ich an einem Beispiel erlautern, das ... aus der Protophysik stammt. Die Aussage 'ein Körper kann seine Geschwindigkeit nur stetig ändern' kann von der modernen Physik nur dynamisch verstanden werden. Geschwindigkeitänderungen sind Beschleunigung, d.h. als Zweite Ableitung des Weges nach der Zeit definiert. Zeit wird von der Physik als ein Parameter behandelt, an dessen Erzeugung durch eine Parametermaschine ("Uhr") de facto bestimmte Homogenitätserwartungen geknüpft sind ... Bezogen auf den Gang einer angeblich so ausgewählten Parametermaschine kann eine Körper seine Geschwindigkeit deshalb nicht unstetig, d.h. mit unendlich große Beschleunigung ändern, weil dazu eine unendlich große Kraft erforderlich wäre" (Janich, 1975:68-69).

who realized that thermodynamic processes are irreversible.[1] Already in 1824 Carnot discovered irreversible processes – a discovery that was elaborated independently from each other in 1850 to the second main law of thermodynamics (the law of non-decreasing entropy). This law accounts for the fundamental irreversibility of natural processes within any *closed system*. The term *entropy* itself was introduced by Clausius only in 1865. In 1852 Thomson explains that according to this law all available energy strive towards uniform dissipation (see Apolin, 1964:440 and Steffens, 1979:140 ff.). Planck remarks that "the irreversibility of natural processes" confronted "the mechanical conception of nature" with "insurmountable problems" (Planck, 1973:55).

It is only on the basis of an insight into the foundational position of the kinematic aspect in respect of the physical aspect that an appropriate designation of the first law of thermodynamics is made possible. Although we are used to employ the familiar designation of it as the law of *energy conservation* there is an element of ambiguity attached to the term 'conservation' – as if energy is "held on to." When, on the law-side, the retrocipation from the physical aspect to the kinematic aspect is captured by the phrase *energy constancy* this ambiguity disappears and then we have at hand a concise and precise formulation of this law.

16 The mystery of matter

The preceding historical sketch made it clear that although each one of the four modes of explanation did open up a legitimate angle of approach none of them can claim to be the *exclusive* and/or *exhaustive* source of our knowledge of material things. Whatever their worth, they merely provide us with a *partial* perspective, one that will always be co-determined by a *totality view* exceeding the scope of any specific mode of explanation. Such a totality perspective actually exceeds the scope of any special science since it inevitably rests upon some or other philosophical view of reality. It entails the necessity to employ modal terms in concept-transcending ways.

These considerations intimately cohere with the modal universality inherent in each modal aspect. Acknowledging the modal universality of the different modal aspects is constitutive for an account of typicality, individuality and concept-transcending knowledge (idea-knowledge).

The impressive power of theoretical thinking fist of all derives from exploring the *modal universality* of specific modal aspects. The philosophically informed physicist Von Weizsäcker implicitly draws upon this insight when he appreciates quantum theory as the *central theory* of contemporary physics. His explanation highlights the modal universality of the physical aspect, for this modal universality is not restricted by the typical nature of any (type of) entity – it cuts across all typical differences. We have noted that Von Weizsäcker says:

---

[1] "Er hat damit das Wesen der im Sinne der Thermodynamik 'nicht umkehrbaren' Vorgänge erkannt. Vom molekular-mechanischen Gesichtspunkte aus gesehen sind dagegen alle Vorgänge umkehrbar" (Einstein, 1959:42).

Quantum theory, formulated sufficiently abstract, is a universal theory for all classes of entities (Von Weizsäcker, 1993:128).[1]

In addition to this appeal to modal universality Von Weizsäcker also explicitly articulates the fundamental philosophical insight that *everything coheres with everything*: ("Alles hängt mit allem zusammen" – Von Weizsäcker, 1993:134).

The modal universality of each one of the four modes of explanation that we have discerned in their successive decisive roles during the history of our understanding of physical nature entails that the scope of each one of them is *unspecified*. This means that whatever concretely exist functions within every one of these modes of existence. We have referred to the law of gravity as an example of the unspecified modal universality of the physical aspect.

But as soon as the typicality of things is acknowledged the unspecified universal meaning of the aspects acquire a *typical specification*, for in this case the *effect* of specific *type laws* is manifested within the modal aspects themselves. In general it can be said that modal laws hold for all *possible classes of entities*, whereas type laws hold for a *limited class* of entities only.Type laws are still universal, but no longer in an *unspecified* sense. For example, although the law for being an atom holds for *all atoms* (its *universality*), it does not hold for all *kinds of entities* (like molecules, planets, animals or states – revealing its specification, for it holds for atoms *only*).

No single entity 'escapes' from having a (typically specified) function within each modal aspect of reality. Naturally this applies to the first four modal aspects in particular. As such this insight already explains why, as we saw from the history of reflection on the nature of material things, every one of them at some or other historical phase was elevated to become the sole explanatory principle of physical reality.

Modal universality and the specificiy of type laws are intimately intertwined. An understanding the former paves the way to an appreciation of the nature of type laws while holding on to the insight that all possible kinds of entities in principle function in all modal aspects. However, merely in terms of the universal scope of the modal aspects it is not yet possible to identify what is *uniquely typical* in respect of any particular kind of entity.

For example, although the formulation of the law of gravity by Newton highlights the universality of the physical aspect, his entire *Principia Mathematica* is still permeated by geometrical language and arguments. This shows its dependence upon the Greek-Medieval space metaphysics preventing him from employing his own calculus. Although the concepts employed by him do take motion and dynamics into account, the general framework continued to be *geometrical* in nature.

In the case of the modal universality of the physical aspect the implication is that every (modal-functional) physical law holds for whatever actively (i.e.

---

1 "Die Quantentheorie, hinreichend abstrakt formuliert, ist eine universale Theorie für alle Gegenstandsklassen."

as a subject) functions in the physical aspect. In connection with the way in which Helmholtz formulated the first main law of thermodynamics – the law of *energy constancy* as we prefer to designate it – Steffens remarks that for "... Helmholtz the law of the conservation of force" includes "all known physical phenomena" (Steffens, 1979:137).

In connection with the question regarding the nature of *matter* this insight leads us towards an understanding of the coherence between the first four modes of explanation.

In our discussion of the earliest phase of the development of Greek thought we alluded to attempts to come to an understanding of what *matter* is. We saw that initially this urge was closely related to the principle of origin sought by some of the prominent pre-Socratic philosophers, mentioning the choices of Thales (water), Anaximines (air), Heraclitus (fire) or the infinite-unbounded chosen by Anaximander. These elements were thought of as flowing, dynamic principles of origin, because at this early stage of Greek thought the motive of form, measure and harmony played a subordinate role. Nonetheless the dialectical struggle between form and the formless was played out in terms of the same basic ontic modes of reality that provided merely a *partial* perspective on material things. Perhaps this is one way to understand the mystery entailed in the question what matter is, for if the aspects as modes of explanation make possible only *partial answers* from functionally distinct angles of approach, then merely making an appeal to these angles of approach will never solve the problem. It is therefore not surprising that Stegmüller believes that one of the most difficult questions facing science in the 20th century is indeed given in the concept of matter, which he considers to be mysterious in the utmost sense.[1]

Of course what Stegmüller has in mind is first of all the old adage that here are no jumps in nature, that it is *continuous* ("*natura non facit saltus*"). Yet we have seen that there is an important difference between physical space and mathematical space – the former is discontinuous since it is bound to the quantum-structure of energy and the latter is continuous and therefore allows for an infinite divisibility. Stegmüller writes:

> In the preceding sections we have repeatedly established how much precisely those disciplines that are focused on the largest bodily structures as such, astronomy, astrophysics and cosmology, remained dependent upon knowledge of the smallest. In fact presently we often cannot even say if the scientific puzzle or theoretical dilemma that here appears poses a challenge merely for the disciplines concerned with the largest or also at once a challenge for the disciplines concerned with matter. One can defend the mean assertion that the contemporary "matter experts" in a certain sense are forced into a worse acknowledgement than Goethe's Faust. They are not only "not wiser than before," namely in respect of the time when they commenced their research, but they

---

1 "Und daß *auf der anderen Seite* ausgerechnet der Materiebegriff der schwierigste, unbewältigste und rätselhafteste Begriff überhaupt für die Wissenschaft dieses Jahrhunderts blieb" (Stegmüller, 1987:90).

simply not once have gotten wiser than those first thinkers who more than 2000 years ago attempted to provide a speculative foundation for matter.[1]

When Stegmüller continues his explanation of the problems attached to an understanding of the nature of matter the first four aspects of reality suddenly acquires a new actuality. In the first place he distinguishes two global basic conceptions regarding the nature of matter and he points out that currently these conceptions once again, as previously, occupy a prominent place in the discussions. He calls these two basic conceptions the *atomistic conception* and the *continuity conception*.[2] Also Laugwitz points out that insofar as physics subjects itself to auxiliary means from mathematics it cannot escape from the polarity between continuity and discreteness.[3]

Suddenly the question concerning the infinite divisibility of matter once again occupies a central position, thus highlighting anew the important distinction between *physical space* and *mathematical space*. It is clear that this distinction between 'atomism' and 'continuity' is based upon number and space as the two most basic modes of explanation of reality. But this is not yet the end of the dependence upon unique modes of explanation, for according to Stegmüller these two conceptions were designed in order to bring to a solution the following two problems (Stegmüller, 1987:91):

(i) The apparent *indestructibility* of matter, and
(ii) The apparent or real limitless transformability of matter.

When these two problems are assessed in their coherence it is immediately clear that they depend upon the third and fourth ontic modes of explanation

---

1 "Wir haben in den früheren Abschnitten mehrmals festgestellt, wie sehr gerade auch diejenigen Wissenschaften, welche sich mit den größten körperlichen Gebilden überhaupt beschäftigen: die Astronomie, die Astrophysik und die Kosmologie, auf das 'Wissen vom Kleinsten' angewiesen bleiben, ja daß wir heute sogar oft nicht einmal sagen können, ob ein hier auftretendes wissenschaftliches Rätsel oder theoretisches Dilemma als bloße Herausforderung der 'Wissenschaften vom Größten' allein aufzufassen ist oder als eine simultane Herausforderung sowohl dieser Wissenschaften *als auch* der Wissenschaften von der Materie. Es ließe sich die boshafte Behauptung verfechten, daß die heutigen 'Materie-Experten' in einem gewissen Sinn zu einem schlimmeren Eingeständnis gezwungen sind als Goethes Faust. Sie sind nicht nur 'nicht klüger als zuvor', nämlich als zu der Zeit, da sie zu forschen anfingen, sondern sie sind nicht einmal klüger geworden als jene ersten Denker, welche vor über 2000 Jahren die Materie rein spekulativ zu ergründen versuchten" (Stegmüller, 1987:91).

2 "Selbst die beiden großen Grundkonzepte über die Natur der Materie stehen heute nach wie vor zur Diskussion, wenn auch mannigfaltig verschleiert hinter Bergen von Formeln. Diese beiden Grundkonzepte kann man als die *atomistische Auffassung* und als die *Kontinuumsauffassung* der Materie bezeichnen" (Stegmüller, 1987:91).

3 "Die Physik, insofern sie sich mathematischer Hilfsmittel bedient oder sich gar der Mathematik unterwirft, kann an der Polarität von Kontinuierlichem und Diskretem nicht vorbei" (Laugitz, 1986:9).

given in reality, namely on the meaning of kinematic persistence ('immutability') and physical changefulness ('transformability').[1]

At this point the key moments in the preceding overview of the history of our understanding of material things recur owing to the fact that all four modes of explanation are still playing a decisive conditioning role in our theoretical reflections. The "thing-ness" of material entities once and for all transcends the limited nature of the unique angles of approach (modes of existence and modes of explanation) that served our understanding of matter. Things function at once within all these modes and yet, in spite of this aspectual many-sidedness of things, their existence is never exhausted by any one of these modal aspects. And it seems that the *mystery* surrounding material entities derives from this multi-aspectual *but-at-once more-than-merely* aspectual nature of such entities.

It is precisely this more-than-merely-aspectual-nature of material things that sheds a negative light on any *monistic* ideal to develop a "theory of everything." With reference to Einstein's search for a unified field theory a specialist in "super string theory" still believes that physicists will find a framework that will combine their insights into a "seamless whole," into a "single theory that, in principle, is capable of describing all phenomena" (Greene, 2003:viii). He indeed presents "super string theory" as the "Unified Theory of Everything" (Greene, 2003:15; cf. also pp.364-370, 385-386). However, he does not realize that although he has a *purely physical theory* in mind, the *meaning* of the physical aspect of reality inherently points beyond itself to its inter-modal coherence with other aspects, first of all with those aspects that are foundational to the physical aspect (namely the aspects of number, space, and movement). Even the way in which he phrases his goal cannot escape from terms that have their original seat within some of these aspects. Just consider his reference to a "seamless whole" and his use of the quantitative meaning of a "single theory," i.e. *one* theory term 'all'. We have repeatedly explained that the core meaning of space (i.e. *continuous extension*) underlies our awareness of *wholeness* and of *coherence* (seamless).

The undeniable interconnectedness of all aspects of reality disqualifies each and every claim to (modal) *purity* – at least when this 'purity' is meant to refer to the meaning of an aspect stripped of its coherence with other aspects.

17 The problem of individuality

Since Greek philosophy a key problem was to find what became known as the *principle of individualization*. Aristotle, for example, claimed that *matter* is the principle of individuality: "But all things many in number have matter" (Aristotle, 2001:884; *Metaf.* 1074 a 34). We have noted above that from the opening sections of Aristotle's work on *categories* it is clear that he took his starting-point in the idea of a *strictly individual* primary substance. Here he

---

[1] The physicist Rollwagen holds the view that the 'dualism' of wave and particle introduced a new dimension, namely the "possibility of the ... mutual transformation of elementary energy-structures" (Rollwagen, 1962:10).

defines substance "in the truest and most definite sense of the word" as "that which is neither predicable of a subject nor present in a subject" (Aristotle, 2001:9; Cat. 2 a 11-13). The secondary substances, however, are universal (Aristotle, 2001:9; Cat. 2-4) – species and genera are universal.

Later on, in neo-Platonism, we find that priority is assigned to what is universal while acknowledging that what is individual *cannot be conceived* (see Plotinus, *Enn.* VI, 3,9,36 and *Enn.* VI, 2,22). Interestingly Simplicius already distinguished between the *numerical one* and what is *individual* (Kobusch, 1976:302).[1] Ammonios Hermeiu influenced the distinction between four "complexions" found in the thought of Boethius: *substantia universalis*; *substantia particularis*; *accidens universale*; *accidens particulare*. Where *singularity* indicates similarity for Richard von St. Victor, individual substantiality is found in one individual only and therefore cannot be shared by multiple substances and for this reason it is 'incommunicable' (see Kobusch, 1976:303).

The commonly held view during the middle ages (shared by Bonaventura, Thomas Aquinas, Henry of Gent and Duns Scotus) was that an individual in itself is *undivided* (the literal meaning of *individuum*) while at the same time it was separated from everything else (see Oeing-Hanoff, 1976:306). The term *undivided* reflects the wholeness (*one*-ness) element of the whole-parts relation and in addition to the 'one' it entails the 'many' separated undivided ones (wholes). In other words, the one and the many are situated within the context of the idea of wholeness and distinctness.

The employment of the whole-parts relation acquired a closer specification in the thought of Boethius who distinguished between *homogeneity* and *heterogeneity* – every part of an individual drop of water is still water (*physical homogeneity*), whereas it is not true that every part of a horse is a horse (*biotic heterogeneity*) (see Oeing-Hanoff, 1976:306).

Leibniz continued the view of Aristotle by inverting the idea that individuality falls under general concepts. Rather one should say that what is universal is contained or embraced in what is particular and individual (see Borsche, 1976:310). When he determines the individual substance in his *Monadology* Leibniz assumes an original self-activity prevailing in a state of *continuous change* (appetition) (Borsche, 1976:311).

In close connection with the early Romantic switch from rationalism to irrationalism Herder affirms that the "deepest foundation of our existence is individual" (Herder, 1877 Vol.II:207). In a letter to Lavater Goethe mentions the saying "Individuality is *ineffabile*" (Borsche, 1976:312). According to Fr. Schlegel individuality is never completed since it is always involved in continuous becoming ("beständiges werden"). What is essentially incompleted is

---

[1] Implicitly this highlights the difference between what we will designate below, from a systematic perspective, as a *conceptual* use of numerical terms (the "numerical one") and a *concept-transcending* use of numerical terms (the idea of an *individual thing*).

infinite and therefore individuality is eternity within the human being and only it can be immortal (Borsche, 1976:315).

Goethe did play with the inseparable connection between individuality and universality. His answer to the question *what is universal?* (*Was ist das Allgemeine?*) is: the individual instance; and his answer to the question *what is particular?* (*Was ist das Besondere?*) is: millions of instances (see Von Weiszäcker, 2002:212). Yet this does not mean that Goethe actually maintained a *balance* between universality and individuality, because according to him, in the words of Von Weiszäcker, the *Gestalt* is not rooted in the law for the law is rooted in the *Gestalt* (Von Weiszäcker, 2002:209).

18 Systematic distinctions

It is striking that the battle-field between *universality* and *individuality* is served by our basic intuitions of *number* and *space*. The idea of *being distinct* (at least partially) pre-supposes the *discrete* meaning of number while understanding *universality* pre-supposes the spatial awareness of *everywhere*. Of course the numerical point of entry – or mode of explanation – can be complemented in yet another way by the spatial angle of approach, namely when its articulation at once also highlights *numerical analogies* within space.[1] This happened during the later middle ages where we have noted that the term *undivided* reflects the *one*-ness element of what we now identify as the *spatial* whole-parts relation.

18.1 Knowledge based upon universality

Although the nature of *individuality* surely exceeds the limits of the numerical (and all other) aspect(s) of reality, it is undoubtedly also the case that in our employment of the idea of individuality our arithmetical intuition is prominent. What is perhaps even more important is that because concepts are formed on the basis of what is *universal* (universal properties), conceptual knowledge of what is *individual* is impossible. Even Scholasticism was faithful to the conviction that whatever is individual is *inexpressible* (omne individuum est ineffabile). However, this limitation of conceptual knowledge led to a reductionist view of knowledge, one in which knowledge is identified with conceptual knowledge. Nontheless we do have knowledge of things in their individuality, the only requirement is to realize that this kind of knowledge transcends the scope of knowledge mediated by universality, i.e. conceptual knowledge. The identification of knowledge with conceptual knowledge may be designated as *rationalistic*.

The kind of knowledge involved in approximating what is *unique, contingent* and *individual* transcends the limits of universality (concepts) and should

---

1 Within the philosophy of Dooyeweerd and Vollenhoven the coherence between the multiple irreducible aspects of reality is accounted for by referring to these interconnections as modal analogies (retrocipations and anticipations). In this context the original numerical meaning of succession (one, another one and so on without an end, endlessly) is turned 'inwards' by the spatial meaning of continuity – seen in the *endless divisibility* of a continuous whole.

therefore be acknowledged for what it is: *concept-transcending knowledge*. Nicolai Hartmann once explained the Kantian notion of a "Grenzbegriff" in a striking way. He says that the notion of an unknowable "thing-in-itself" ("Ding an sich") still requires a *thought-form* through which it is thought of *as* unknowable – and this is what a "Grenzbegriff" intends to capture.[1] Without buying into the role of the so-called "thing-in-itself" in the philosophy of Kant (cf. the critical remarks made in Strauss, 1982:133, 141-143), it is important to leave room for a "form-of-thinking" accounting for knowledge transcending the limits of concept-formation.

In his work on logic and epistemology De Vleeschauwer explicitly mentions what he calls the "individual delimitation" (De Vleeschauwer, 1952: 213). He writes that the domain of the "individual" is one where our intellectual capacities must fail. His own *nominalistic* affinities are evident in his view that there are only *individual things* and processes. He holds that in spite of all similarities between entities and processes, there will always remain an irreducible kernel of *individuality*, which causes one thing to be different from another one. Science with its directedness towards the *universal* has serious difficulties with its inclination also to know what is individual – because "knowledge of what is individual is simply impossible" – something about which philosophy, according to De Vleeschauwer, had clarity since its inception (De Vleeschauwer, 1952:213). In other words, De Vleeschauwer adheres both to the nominalistic denial of universality outside the human mind and the rationalistic identification of knowledge with conceptual knowledge.

18.2 *Knowledge exceeding universality*

From what has been argued thus far it is clear that the relation between *law* and *individuality* at least in one sense runs parallel with the distinction between *universality* and *individuality*. Whereas we have pointed out that the notion of universality cannot conceal that it is derived from the meaning of the spatial aspect – the awareness of *everywhere* – the situation with knowledge of what is individual is more complicated.

We argued that knowledge of what is individual exceeds the confines of conceptual knowledge. Yet we can specify what is here at stake in more precise terms. Consider for a moment how we can apply our basic intuition of number, space, movement and energy operation in the following case. Think of the *quantitative* properties of a cultural entity (like a chair), of its *size and shape* (spatial), of its relative speed (*motion*) and of its typical physical characteristics (it *strength*). In every instance the terms that we have employed inevitably have a *universal scope*. This means that whenever any person looks through the gateway of these different (ontic) points of entry[2] at a chair, the

---

1 Cf. Hartmann, 1957:311: "Denn bei Kant ist es nicht so, dass etwa das Ding an sich bloss Idee wäre; umgekehrt, da wir das Ding an sich nicht erkennen ..., wohl aber denken können, so muss es eine Denkform, eine Art des Begriffs geben in der es – eben als unerkennbares – gedacht wird. Das ist die 'Idee'."

2 Which are then at once elevated to serve as epistemic modes of explanation.

terms generated are used in a *conceptual* way. As long as we restrict the use of such terms to the respective ontic domains (modes of explanation) this conceptual focus will always be present. This is actually the case with all our entitary-oriented everyday concepts (just think of our concepts of entities such as planets, houses, chairs and human beings). If we designate the terms employed in describing the way in which entities function within various aspects of reality as *modal terms* (see Strauss, 2000:26-28, 32-36), then the following distinction ought to be drawn. When modal terms are used to refer to entities that function within the confines of particular modes of being, they are employed in a *conceptual* manner. However, whenever a modal term is put in service of referring to whatever exceeds the limits or boundaries of such an ontic domain, then we encounter a *concept-transcending* use of such a term – also designated as an *idea-use* of such terms.

For example, while merely exploring our quantitative intuition, one can speak of a chair in its totality, including all its properties. Linguistically this is expressed by referring to its *individuality*, its *uniqueness*, its *being distinct*. The original quantitative meaning of number (discrete quantity) – captured as a "primitive term" in axiomatic set theory[1] – is evident in these affirmations and yet they are intended to refer to much more than merely the arithmetical aspect of the chair. They therefore indeed constitute idea-usages of modal numerical terms.

Similarly, instead of speaking of the sizes and dimensions of a chair, one may use our intuition of the original meaning of spatial extension to speak of all facets of the chair – in which case one may refer to the chair in its *totality*. Once again it is clear that the term totality – in spite of its spatial descent – here refers to much more than merely the spatial aspect of the chair. It constitutes therefore – in terms of the distinction suggested by us concerning the twofold usage of modal terms – another example of an idea-use of such terms, in this case spatial ones.

Modern phoronomy (the pure science of movement) understands motion in its original sense as uniform flow, without the need of any causes (as Aristotle believed). This kinematic intuition of constancy, when used in an idea-context, provides us with the idea-knowledge of the identity of an entity – its relative constancy amidst all changes – where the latter term finds its seat in the physical aspect of energy operation. The operation of energy always causes certain effects and in that sense never leaves anything the same, i.e., identical. Therefore, the word change can also be employed in an idea-context. But because the idea-meaning of constancy (consonant with the idea of *identity*) and the idea-use of the term *change* stem from two irreducible modes (detecting changes always presuppose constancy), it is not contradictory to use both these ideas concurrently.

---

[1] Given in the plural of "elements of" or, in the case of Zermelo Fraenkel set theory, "members of."

By expanding our view we can indeed highlight the four most basic idea-statements philosophy can formulate about the universe – and once again we have to realize that these statements are not contradicting each other but rather entail and complement each other: (i) everything is unique; (ii) everything coheres with everything else; (iii) everything remains identical to itself; and (iv) everything changes. Only if these statements did not rest upon irreducible modal points of entry they would have been contradictory.

At this point we may unite the main contours of our historical overview by pointing out that most of the issues mentioned converge towards an understanding of the role of the four aspects underlying the last mentioned four idea-statements about reality. In fact we have seen that these aspects served as points of entry in the history of reflection on law and individuality. Ultimately no view on law and individuality can side-step the first four modes of explanation of the world – for in both instances one encounters a mixture of conceptual terms and a concept-transcending usage of modal terms. In order to articulate this claim we now focus on the nature of *natural laws*.

18.3 The concept of a normative principle and a natural law

Cassirer mentions another element present in the ancient Greek understanding of *law* – namely that *nomos* constitutes a principle of *ordering* through which motion and the diversity within reality is arranged (Cassirer, 1911:375). During the early modern period a reaction to the traditional Aristotelian-Thomistic view led to a natural scientific orientation that treated law predominantly within a *relational coherence*.

This new accent is an effect of a fundamental switch, one in which the focus is not any longer on the *substance* of things (their *essence*), but merely on the *way* in which we experience them. Galileo is therefore no longer interested in the 'essence' of things but instead asks *how* they appear to us. What is revolutionary in his view, according to Herold, is that in the absence of thinking about essences (that proceeded from configurations of motion with distinct degrees of perfection) everything in principle is equal in the face of the law – amply demonstrated in his remark that he did not study the pedigree of geometrical figures (see Herold, 1974:502).

Up to this point the following features of natural laws surfaced: (i) that they are *necessary*,[1] (ii) they constitute an *ordering* in the sense of a *relational coherence* regulating *motion* and *diversity* within reality, and (iii) they are concerned with the *how* and not the concrete *what* of things.

During the early modern era these ideas developed within the context of the dominance of the modern ideal to understand the universe in terms of the (mathematical) natural sciences (also known as the natural science ideal). In the thought of Hobbes science, understood as (natural) philosophy, opened the way to view individual things in relation to what is *universal*. His empha-

---

1 Stegmüller considers it to be "Hume's great achievement to have banished the concept of necessity from the concept of cause" (Stegmüller, 1977:36).

sis on the *truth* entailed in universal propositions (see Herold, 1984:503) reveals his nominalistic affinity that projected a human element into the universality of a natural law. Particularly Newton (in his *Principia Mathematica*, 1, 15) started to explore more extensively the view that a law must be understood as a *mathematically conceivable rule*. While taking distance from the idea of a God-given law the French Enlightenment, particularly D'Alembert, derived from the relations between bodies governed by law the *validity* of the latter (Herold, 1974:505).

In the thought of Kant the feature of necessity (*Notwendigkeit*) is accompanied by what he claims to be the *universal-validity* of a law. Insofar as rules are objective they are designated as *laws*. We have seen that Kant derived these laws in an a priori way from human understanding that furnishes phenomena with the law to which they are subjected, i.e. understanding creates their lawfulness or law-conformity (*Gesetzmäßigkeit*) (see also Kant, 1781-A:126). Kant's aim is to render comprehensible the "objective validity of the pure concepts a priori" of the categories of understanding (Kant, 1781-A:128).

Hegel explored a further dimension in his science of logic when he focuses on determining law (*Gesetz*) as what remains the *same* in what *changes* (Hegel, 1957-2:122). Cassirer assumes "ultimate logical invariants" that are not affected by their changing material content. He speaks of "identity and permanence" that lie "at the basis of scientific laws" (Cassirer, 1953:325). He actually got quite close to an understanding of the conditioning role of the first four modal aspects of reality in connection with an articulation of the nature of natural laws:

There is no objectivity outside of the frame of number and magnitude, permanence and change, causality and interaction: all these determinations are only the ultimate invariants of experience itself, and thus of all reality, that can be established in it and by it (Cassirer, 1953:309).

The "ultimate invariants of experience itself" are actually referring to the conditioning role of the most basic modal aspects of reality – they are indeed those 'determinations' co-responsible for the way in which we experience reality.

Without acknowledging the ontic structural configuration of reality and in particular the ontic order of successive modal functions, we are left afloat without an anchoring guideline in our attempts to define the nature of natural laws. But once the underlying and conditioning role of these (ontic) modes of reality is acknowledged another challenge faces theoretical analysis, namely to distinguish between (i) elementary and (ii) compound basic concepts.

(i) Every scholarly discipline that finds its field of investigation delimited by a specific aspect of reality employ basic concepts reflecting the inter-connections between its delimiting aspect and other aspects of reality. In the case of physics, for example, its elementary basic concepts articulate the analogies of pre-physical aspects within the structure of the physical aspect, found in phrases such as physical mass (numerical analogy), physical field (spatial), and energy-constancy (the kinematical).

(ii) On the basis of elementary (analogical) basic concepts, successively open for theoretical inspection and analysis, the compound (or complex) basic concepts of a scientific discipline are formed by simultaneously involving distinct analogical basic concepts.

Without exploring the methodology of compound basic concepts in any detail we may briefly state the result of such an analysis with respect to the nature of *norming laws*, i.e. of the normative principles guiding human action, such as logical principles, historical principles social principles, aesthetic principles, jural principles, and so on. Mainly in order to side-step the inherent problems of traditional views of "natural law" (as a universally-valid system of law founded in human reason and holding wherever for all times *per se*) an account of the nature of a principle ought to distinguish between a principle given as mere starting-point for human action and the diverse, historically changing ways in which such a point of departure could be given a positive shape in specific unique circumstances.

Yet natural law did see something of the underlying (universal, constant) structure of our legal experience, but it distorted its meaning by assuming that those underlying principles are already (for all times and all places) *made valid* (*enforced*). No principle in this fundamental ontic sense is valid *per se*. Every principle requires human intervention in order to be made valid, i.e. no (pre-positive) ontic principle holds by and of itself. Only human beings are capable to 'enforce' them (as Derrida correctly emphasizes),[1] and only human beings can give a positive form or shape to them. The activity of giving form to underlying principles is sometimes designated as acts of positivizing and the result of such acts are accordingly also known as positivizations. Habermas frequently speaks of "positivizing law" ("die Positivierung des Rechts") (see Habermas, 1996:71 and 1998:101, 173, 180). It is this intermediate and dependent position of all positivizations that is reified by the idea of the (logical, lingual or social) *construction* of the world.

The most basic way to characterize the pre-positive nature of a principle is therefore given in the exploration of the point of entry of the first three modal aspects of reality, namely when we say that a principle is a *universal* and *constant point of departure* for human action.[2] Once this has been said one can proceed by saying that such a principle can only be *made valid* (enforced) through human intervention, i.e. through the action of a competent organ with an accountable free will enabling a proper interpretation of the unique circum-

---

1 Derrida says that there "are a certain number of idiomatic expressions" in the English language that "have no strict equivalent in French," such as the phrase "to enforce the law," or the phrase "the enforceability of the law" (Derrida, 2002:232).

2 The term *universal*, as we have argued, derives from the spatial awareness of *everywhere*, the term *constant* from the core kinematic meaning of *uniform flow* and the phrase *point of departure* underscores the *unity* and *distinctness* of each principle.

stances in which the principle should be given a concrete shape (should be positivized).[1]

The account of a pre-positive principle overlaps with the constitutive elements that ought to be incorporated in an account of natural (physical) laws. Van Fraassen refers to Pierce who argues that *if* the 'uniformity' intended by Mill merely meant *regularity* without any real connection implied between events, then his argument will be destroyed (Van Fraassen, 1991:22). The phrase used by Van Fraassen in this context, however, states that a law cannot be "the mere uniformity or regularity itself" for a "law must be conceived as the reason which accounts for the uniformity in nature" (Van Fraassen, 1991:22). The use of the word 'reason' may be interpreted to suggest that laws result from the intellectual endeavors of human beings. Nonetheless he continues by claiming that a "law must be conceived as something real, some element or aspect of reality quite independent of our thinking or theorizing – not merely a principle in our preferred science or humanly imposed taxonomy" (Van Fraassen, 1991:22-23). Within the above-mentioned context this implies that the word 'reason' rather means 'cause' – in the sense that a law is the (extra-mental) cause that accounts for the uniformity or regularity of nature. Of course a much easier account would be to state that the regularity of nature concerns its orderliness or law-conformity, entailing that whatever behaves in law-conformative ways is *subject* to a law as *order for*. For that reason perhaps the best translation of the German term *Gesetzmässigkeit* (Dutch and Afrikaans: *wetmatig*) is *subject to law* (as it was done in the translation of Wittgenstein's *Tractatus*).

When Van Fraassen discusses the views of Davidson he points out that although Davidson does not attempt to *define* laws it is nonetheless said "that laws are general statements which are confirmed by their instances" (Van Fraassen, 1991:33). In this case the distinction between *ontic laws* and human statements intended to capture conceptually what such laws are all about collapses. The acceptance of ontic laws does need 'markers', i.e. terms helping us to articulate their ontic nature. The mere fact that we speak in the *plural* about such laws already reflects the constitutive role of the meaning of number (the *one* and the *many*) in our understanding of ontic laws. Furthermore, without the conditioning role of the spatial aspect it cannot be asserted that laws hold *everywhere*, i.e. that they apply *universally*. Although Van Fraassen does acknowledge *universality* as a "mark of lawhood" (Van Fraassen, 1991:26) he does mention with reference to Armstrong and Lewis that the "criterion of universality" is "no longer paramount" in a "discussion of laws" (Van Fraassen, 1991:28).[2]

---

1   Once again the key terms employed in this formulation are derived from diverse modal aspects: 'organ' from the biotic; 'will' (desire) from the sensitive; 'accountable' from the logical-analytical, 'giving shape' (positive form) from the cultural-historical, and 'interpretation' from the lingual modality.

2   This may be the effect of not distinguishing between modal laws (with an unspepcified universality) and type-laws (with a specified universality – see page 56 below).

Of fundamental importance is the distinction between *modal laws* and *type-laws* for it entails an account of the difference between *unspecified* and *specified* universality. A modal physical law holds for *all* kinds of physical entities without any specification, whereas a physical type-law only holds for a limited class of entities, namely those belonging to that type. Such a type-law, for example the law for being an atom, holds universally in the sense that it applies to all atoms, but this universality is *specified* since it holds for atoms only (and not for every kind of physical entity).

In the case of modal laws and type-laws the reverse side of universality is found in the *distinctness* of different laws, specified by using the idea of *delimitation* that is derived from the primitive meaning of space. Every law has its own *domain* of application, a specific and distinct *sphere* within which it obtains. But only when these two elements are combined with the *constancy* (or *uniformity*) of a law and with its *effect* (its *force*, its *validity*) is it possible to account for the constitutive elements of the compound basic concept of a natural (physical) law.

When Stegmüller discusses the law of causality he introduces multiple terms closely related to the conditioning role of the first four modal aspects and in fact approximates closely our idea of *compound basic concepts*.

> Still others might be added to the features we have already mentioned. But since with these the concept of a causal law has already reached such a high level of complexity, let us confine ourselves to them. For faced with the question 'what are causal laws?' we must, in accordance with them, give the following answer: *causal laws are quantitative, deterministic, nearby action, succession micro-laws formulated by means of continuous mathematical functions in relation to a homogeneous, isotropic spatio-temporal continuum governed by certain principles of conservation* (Stegmüller, 1977:36).

Dooyeweerd did not apply his own transcendental-empirical method of analysis to the idea of 'law' in general, for he frequently simply (intuitively) states that a law *determines* and *delimits* whatever is subjected to it (see Dooyeweerd, 1997-I:508).

If the distinct *scope* of laws delimit their unique areas of validity, then it is recommendable not to allow the concept of a natural law to degenerate into an amorphous collection of predicates, such as it is found in the recent proposal of Stafleu. He says that a law is sometimes hidden behind the name axiom, constant, proposition, rule, relation, thesis, symmetry, theorem, design, pattern, connection, prohibition, comparison, phenomenon or prescription (see Stafleu, 2002:39). This list contains elements referring to the law-side and the factual side as well as a mixture of ontic phenomena and what is the product of human activity. For example, for the sake of convenience Stafleu calls a mathematical law-conformity ("wiskundige wetmatigheid") such as the theorem of Pyhtagoras a natural law (Stafleu, 2002:39).

Inherent to a natural law is its meaning as an *order for* – and this mode of speech makes an appeal to the *unity in the multiplicity* of different laws, for without such a *unity* laws will clash and will not be able to constitute an *order of laws* (a law-order). The constitutive role of the numerical mode is evident in this concept of *order*. Furthermore, a law entails its *correlate*, namely that which is factually subjected to it – and this insight points at the inherent *universal scope* of a law – derived from the spatial awareness of *everywhere* (at all places). Without the spatial (dimensional) distinction between *above* and *below* the assumed correlation of *law* and *subject* does not make any sense. That the *validity* of a natural law is not something incidental is captured by saying that it holds *constantly* – demonstrating the constitutive role of the kinematic mode. The notion of *validity* (*being in force*) derives from the core meaning of the physical aspect and it has to be incorporated in the concept of a physical (natural) law, because otherwise the ability to say that a law *determines* what is subjected to it would collapse.

The compound or complex basic concept of a natural law may therefore be formulated as follows;

> As a unique, distinct, and universally valid order for what is factually correlated with and subjected to it, a natural law constantly holds (either in an unspecified way as in the case of modal laws or in a specified way as in the case of type laws) within its domain of validity.

However, what is particularly striking in reflections on the relation between *law* and *factuality* is the widely found confusion of *law* and *law-conformity*.

19 Physical entities exceed the limits of physics

Our analysis of the development of the theoretical understanding of physical reality made it clear that the to find an *arithmetic, spatial* or *kinematic* qualification for *physical entities* necessarily runs into theoretical antinomies. Although material entities without any doubt do function within these three (pre-physical) aspects, they are still always qualified by the *physical* aspect. Consider for a moment the many-sided existence of an atom.

Already in 1911, in Rutherford's atomic theory, the hypothesis was posed that atoms consist of a (electrically positive) nucleus and negatively charged particles moving around it (a view which was inspired by the nature of a planetary system). In the following year (1912), Niels Bohr set up a new theory which contained two important new ideas: (i) the electrons move only in a limited number of discrete orbits around the nucleus and (ii) when an electron moves from an orbit with a high energy content to one with a low energy content, electromagnetic radiation occurs. In 1925, Pauli formulated his exclusion principle (Pauli-exclusion).[1] According to the division of charges of electrons, corresponding electron-shells exist, and in each peel there is room for a 'maximum' number of electrons. This maximum number is given by the simple formula: $2n^2$. In the first peel (known as the K-peel) there is room for 2

---

[1] It applies to fermions, i.e. elementary particles with a semi-integral spin (1/2, 3/2, 5/2, etc.) for which the statistical laws of Fermi-Dirac are formulated.

electrons; in the following L-peel, there is room for 8; in the M-sheel for 18; in the N-sheel for 32; and so on. Within a sheel with a quantum number n, (where there is room for $2n^2$ electrons) sub-orbits are identified so that each sub-orbit with a quantum number l has room for $2(2l+1)$ electrons.

Multiple elementary particles are integrated in the unified functioning of the atom as an individual *whole*. When physicists talk of the nature of these particles the original meaning of space, combined with numerical analogies within space, is prominent. In other words, the aspects of number and space are first explored in what we say about elementary particles.

The nucleus of the atom (constituted by its protons and neutrons) has a certain *size* and its diameter multiplied by 100 000 specifies the distance between the nucleus of an atom and its (circling) electrons. The current physical view is that quarks are the ultimate "building blocks" of these elementary particles.[1]

The matter of an atom is concentrated in a volume of less than a 0,0000000000000000000001 part of the volume of the atom – which amounts to saying that atoms are for more than 99,9999999999999999999% empty (19 zeros after the comma – Kiontke, 2006:27). Some facts about the way in which atoms function within the kinematic aspect are equally astonishing. According to wave mechanics we find quantified wave movements around the atom and the electron of a hydrogen atom (in its lowest orbit) moves around the nucleus at a speed of about 6.8 million km per hour (Kiontke, 2006:27).

From these facts it is eviden that the *distinct number* of elementary particles within the internal atomic structure are joined into a typical *spatial* and *kinematic* order of electronic orbits that configure the atom as an individual *physical-chemical micro-totality*. The special spatial configuration which is manifest within the internal arrangement of the parts of an atom reflects the typical *foundational function* of atoms. Biochemistry discovered many *isomeric* forms, that is, they have identified *chemical configurations* that are constituted by the *same* atoms, viewed from a purely numerical perspective, but that nonetheless, owing to different *spatial* arrangements, differ *chemically*. The formula C_3H_6O may yield the following (chemically distinct) configurations: $CH_3.CH_2.CHO$ or $CH_3.CO.CH_3$. Another example is $C_4H_4O_4$. That the chemical differ-

Maleic acid (cis)

Fumaric acid (trans)

---

[1] A distinction is drawn between an *up quark* (with a charge of +2/3) and a *down quark* (with a charge of -1/3) – but apparently there are not any *free* quarks. The proton, for example, consists of two up quarks and one down quark. The *size* of electrons and quarks is smaller than 10^{-18} – they are so *small* that they are described as *point-like* (see Kiontke, 2006:27). *Hydrons* include those *fermions* and those *bosons* designated as *mesons*. Furthermore, hadrons are constituted by *quarks*. Those known as *baryons* in turn include nucleons (neutrons and protons) and *hyperons*. Whereas the hadrons are 'heavy' the *leptons* are *small*, including the electron and particles such as the *muon*, *tauton* and their corresponding *neutrinos*. More information on this micro-dimension is found in Penrose (2005:645 ff).

ence between maleic acid and fumaric acid has its foundation in alternative spatial arrangements is self-evident – just as clear as it is that the number of atoms as such cannot account for this chemical difference. In other words, it is intuitively clear that molecules such as these have a spatial foundational function and not a numerical foundational function. The point-like nature of small particles may suggest that in their case the *numerical* function is decisive. One may be tempted to argue that such particles have their foundational function in the numerical aspect, but the problem is that in the case of electrons and quarks their actual size is unknown (Kiontke, 2006:27). Nonetheless, as Stafleu remarks, the electron is characterized by exactly determinable values for its charge, rest mass, the magnetic moment and the lepton number (Stafleu, 1989:91).[1]

The problem of the interweaving of different kinds of entities first of all challenges the limitations of the whole-parts relation – a relation that appears in its original modal meaning within the spatial aspect.[2] Suppose we ask whether or not Sodium and Chlorine are genuine *parts* of table salt. Surely every division of table salt must continue to display the NaCl structure of table salt. But what happens when the process of division reaches a single salt molecule? Once such a molecule is divided one ends up with a Sodium atom and a Chlorine atom – and it is evident that real parts of salt will still possess the same chemical structure of salt, namely NaCl. The critical question is whether Sodium and Chlorine each has a salt structure, i.e. are Sodium and Chlorine true parts of salt? The answer is self-evident because neither one on its own has a NaCl structure!

In this case the internal sphere of operation of the atoms remained intact althought, through the chemical bond, they were taken up in the talble salt molecule. Dooyeweerd to develop a theoretical approach in order to account for the retention of the internal nature of entities that are interlaced (cf. Dooyeweerd, 1997-lll:627 ff., 694 ff.). When the internal sphere of operation of a interwoven entities is retained, the term *enkapsis* is employed. When one kind of entity is foundational for another kind of entity, the situation is captured by speaking of a *one-sided enkaptic foundational relationship* – which is what we found in NaCl.

1 A detailed explanation of primary and secondary foundational relations is found in Stafleu, where, for example, *energy, force* and *current* respectively are related to *quantitative, spatial* and *kinematic* concepts (Stafleu, 2002:26-28, 128-171). His analysis explores the possibility to discern the first three modal aspects as the foundational function of different kinds of physical entities – and in doing that demonstrates how fruitful philosophical distinctions are for a special science such as physics. Quantum electro-dynamics has to take into account the interaction of the electron with its own quantified surrounding field – charge and field are inseparably connected (see Rollwagen, 1962:10).

2 Note the difference between mathematical space (that is continuous and infinitely divisible) and physical space (that is discontinuous and therefore not infinitely divisible). Both kinds of space are *extended* – the similarity between them; but within the moment of similarity the difference between them manifests itself, thus demonstrating the nature of a (modal) analogy.

Within the realm of physically qualified entities we encounter different geno-types. Atoms are, for instance, geno-types within the radical type (realm) of material things. Within different bonds the same atom displays a number of *variability types*. When an atom engages in chemical bonding, a characteristic enkaptic totality emerges: (i) besides the internal sphere of operation of an entity there is (ii) an external enkaptic sphere of operation in which the enkaptically-bound entity serves the encompassing enkaptic totality.

The factual configuration of a water molecule thus exists on the foundation of the geno-type of the chemical bond between the oxygen and hydrogen atoms. Without these atoms a water molecule cannot exist. They serve water therefore in the sense of a unilateral foundational relation. Does this imply that the atoms totally become part of the chemical bond that exists within the molecule? Not at all, because the bond applies only to the binding electrons and not to the whole atom. Besides, the atom nucleus is not just a specific characteristic of the atom, but precisely that nuclear part of an atom that determines its physical-chemical geno-type (compare the atomic number = the number of protons of the nucleus), as well as the atom's place within the periodic table.

The fact that the atom nucleus remains structurally unchanged in the chemical bonding guarantees the internal sphere of operation of the atom. Because the electrons cannot be disengaged from the nucleus of the atom these atoms function as a whole in the water molecule. Note that it cannot be said that the atoms function in the chemical bond for the bonding does not encompass the atomic nuclei. Nonetheless the atoms (with their nuclei, electron shells and bonding electrons) are present *as a whole* in the water molecule which embraces them *enkaptically*. The enkaptic interweaving of the atoms within the molecule does not make them intrinsical parts of the molecule, since this would abrogate the internal sphere of action of the atoms.

The external enkaptic function of the oxygen and hydrogen atoms in the water molecule indicates the functioning of the atoms within the molecule as a totality via the intermediate role of the chemical bond. This presents us with a three-fold distinction:
(i) First of all, we must identify the internal sphere of action of the atom.
(ii) Secondly, we find the chemical bond that leaves the atom nucleus unchanged because it only affects the outer electron shells, so that the atom nuclei can in no way be part of the chemical bonding.
(iii) Thirdly, we find the enkaptic structural whole of the water molecule that enkaptically embraces the atomic nuclei and bonds and ascribes to each of them their typical place within the enkaptic whole.

This view side-steps the one-sidedness of an atomistic understanding of a molecule (over-emphasizing the continued existence of atoms within the chemical bond at the cost of the totality character of the resulting whole) and of a holistic view (that over-emphasizes the totality-character of the molecule in such a way that the foundational atoms are seen as *integral parts* of the whole). Van Melsen highlights these two extremes: "In modern theories atomic and molecular structures are characterized as associations of many interacting entities that *lose* their own identity. The resulting aggregate originates from the converging contributions of all is components. Yet, it forms a new entity, which in its turn controls the behaviour of its components" (Van Melsen, 1975:349).

It should be noted that also in this context there is an ambiguity in Dooyeweerd's terminology, because he talks of the interlacement of *individuality-structures* (i.e. of *laws*) – instead of clearly stating that the issue concerns the interwovenness of *entities* subject to their type laws (individuality-structures). The intention is not to say that *laws* are enkaptically intertwined, but simply to account for the factual interlacements found between different kinds (types) of entities in their subjection to type laws.[1]

20 Concluding remark

Our preceding analyses are informed by the radical and integral biblical ground-motive of creation, fall and redemption, articulated in terms of a philosophical understanding of reality that in principle takes position against all forms of reductionism. The implications of the non-reductionist ontology explored in our exposition were directed at the uniqueness and mutual coherence of the various aspects of reality, in particular the first four modal aspects: number, space, the kinematic and the physical.

Such an approach is critical of "theories of everything." Breuer holds that the idea of *reductionism* provided the most important stimulus for engaging in "theories about everything" – but his own view approximates the idea of sphere-sovereignty. If one does not assume reductionism it is "possible to advance a universally valid physical theory, another universally valid biological theory and in addition a universally valid economic theory" (Breuer, 1997:3).

The apparently "most exact" of all the sciences, mathematics and physics, cannot liberate those working within these scholarly domains from a theoretical view of reality exceeding the confines of their respective fields of investigation and therefore in principle only have two options: (i) explore the implications of a non-reductionist ontology for these disciplines or (ii) succumb to reductionist thought-patterns with their antinomous consequences.[2]

---

[1] Stafleu also speaks of the mutual interlacement of *laws* ('character's' in his terminology, as clusters of laws) (see Stafleu, 2002:150).

[2] See Appendix II where a brief indication is given of a multi-aspectual perspective on the wave-particle duality.

Appendix I

The 'exactness' of a mathematical proof dependent upon philosophical assumptions

1 Non-denumerability: Cantor's Diagonal Proof

A set is called (d)enumerable when its elements can be correlated one-to-one with those of the set of natural numbers, i.e. any set the elements of which can be arranged in a natural sequence of 1, 2, 3, 4, 5, 6, ... It is clear that the integers are *denumerable*: 0, -1, +1, -2, +2, -3, +3, ... Since all rational numbers can be depicted by two integers in the form of a/b (with b ≠ 0), it is clear that they also can be denumerated. Notice the course of the arrows in the following depiction:

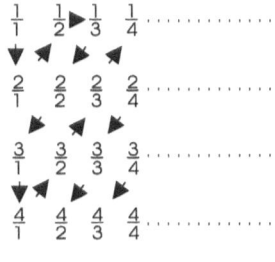

Even all algebraic numbers are denumerable.[1] In a letter of 29 November 1873 Dedekind mentions to Cantor that he had proven that all algebraic numbers are denumerable (cf. Meschkowski, 1972:23). Dedekind does this by defining the height h of an algebraic number x satisfying a polinomial equation $a_n x^n + a_{n-1} x^{n-1} + ... + a_1 x + a_0 = 0$ as follows:

$$h = n - 1 + |a_0| + |a_1| + ... + |a_n|$$

Since the coefficients a_n are integers, only a finite number of algebraic numbers belong to each height h. Since every finite quantity is denumerable, the algebraic numbers as such are also denumerable (cf. Meschkowski, 1972:24).

---

1 Algebraic numbers are the roots of algebraic equations.

Appendix

In 1874 however Cantor proved that the real numbers are not denumerable (i.e. are non-denumerable). Only in 1890 does he provide his diagonal-proof, which we use in our explanation below (cf. Cantor, 1962:278-281). A one-to-one correspondence could be established between all real numbers and the set of real numbers between 0 and 1. Furthermore, every real number in this interval can be represented as an infinite decimal fraction of the form $x_n = 0.a_1a_2a_3a_4$ ... (numbers with two decimal representations, e.g. 0.100000... and 0.099999 ... are consistently represented in the form with nines). Suppose a denumeration $x_1, x_2, x_3,$ ... exists of all the real numbers between 0 and 1, i.e. of all the real numbers in the interval $0 \leq x_n \leq 1$ (i.e. [0,1]), namely:

$x_1 = 0.a_1\ a_2\ a_3...$ ....

$x_2 = 0.b_1\ b_2\ b_3$ ....

$x_3 = 0.c_1\ c_2\ c_3$ ....

....

If *another* number can be found between 0 and 1 which differs from *every* x_n, it would mean that every denumeration of the real numbers would leave out at least *one* real number, which would prove that the real numbers are non-denumerable. *Such* a number we can construe as follows:

y = y₁y₂y₃y₄ ..., with $y_1 \neq 0$, a_1 and 9; $y_2 \neq 0$, b_2 and 9; $y_3 \neq 0$, c_3 and 9; and so forth.

It is clear that y is a real number between 0 and 1 (i.e. $0 \leq y \leq 1$). The number y does not have two decimal representations since every decimal number in its decimal development is unequal to 0 and 9. The number is also unequal to *every* real number x_n since the decimal development of y in the *first* decimal place differs from the *first* decimal number x_1, in the *second* differs from the *second* decimal number of x_2 (namely x_2), and in general from the *n*th decimal number of x_n. It is clear from this that a denumeration of the real numbers will always exclude at least *one* real number ("miscount" it in the denumeration), which concludes Cantor's proof that real numbers are *non-denumerable*.

Also Brouwer (who identifies existence and constructibility and denies the validity of the logical principle of the excluded third with regard to the infinite) rejects the completed infinite.[1] With this Brouwer rejects the transfinite arithmetic of Cantor. The notion of *countability* after all only becomes particularly relevant after the demonstration of the existence of non-denumerable

---

1 "immers de intuitionist kan geen andere, dan aftelbare wiskundige verzamelingen construeeren" (1919:24).

cardinalities. Did Cantor not demonstrate that the set of real numbers is non-denumerable?

In Cantor's diagonal proof it is assumed that all, i.e. the at once infinite set of real numbers, are correlated one-to-one with the set of natural numbers, after which it is demonstrated that a further real number can be specified that differs from each of the *counted* real numbers (in at least one decimal place), from which the non-denumerability of the real numbers is concluded. The validity of this conclusion depends, however, on the acceptance of the completed infinite. Someone who recognizes only the potential infinite can never accept this conclusion, since the diagonal method then only proves that for a given constructible *sequence* of *countable* sequences (i.e. decimal expansions of real numbers) of natural numbers, yet another different countable sequence of natural numbers can be construed. Becker states this in the following way: "The diagonal method demonstrates, strictly speaking, the following: when one has a *counted* (law-conformative) sequence of successive numbers, a sequence of successive numbers can be calculated which differs in every place from all the previous ones" (Becker, 1973:161 footnote 2). In this interpretation there is no room for non-denumerability!

A mathematical proof which apparently takes an 'exact' course therefore results in conflicting conclusions depending on the presuppositions (namely *completed infinity* or *uncompleted infinity*) from which one proceeds! Fraenkel points this out emphatically:

> Cantor's diagonal method does not become meaningless from this point of view, ... the continuum (i.e. the real numbers – DFMS) appears according to it as a set of which only a countable infinite subset can be indicated, and this by means of pre-determinable constructions (Fraenkel, 1928:239 footnote 1).

Whoever rejects the actual infinite (the at once infinite) cannot accept the description of real numbers given by Dedekind, Weierstrass, and Cantor.

Appendix II

The wave-particle duality

1 Complementarity – limits to experimentation

There are also remarkable *limits* to physics in the sense of experimental exactitude and determination. By introducing his principle of uncertainty Heisenberg showed that it is impossible simultaneously to measure the *impulse* and *position* of an electron. The Copenhagen interpretation of quantum physics employs the notion of *complementarity* in order to account for the impossibility to establish both at once – thus allowing for two irreducible (and complementary) modes of description, in terms of "place" and "impulse" respectively. In following some ideas of Mario Bunge the physicist Henry Margenau defends a so-called "moderate reductionism." He takes this the be "the strategy consisting of reducing whatever can be reduced without however ignoring emergence or persisting in reducing the irreducible.[1]

2 The typical totality structure of an entity (wave and particle)

After Einstein reverted to a particle theory regarding the nature of light,[2] it turned out, on the basis of *interference phenomena*,[3] that it is always possible to ascribe a wave-character to elementary particles. Conversely, the *Compton-effect* – regarding the interaction of a photon and an electron – supplied evidence to support the idea of *distinct* particles. De Broglie broadened the perspective by showing that with each and every moving particle (atoms, molecules and even macro-structures) one can associate a wave (cf. Eisberg, 1961: 81, 151).

Although it turned out to be impossible to establish experimentally at *the same time* both the particle and the wave nature Bohr claims that these two perspectives are *complementary* (cf. Bohr, 1968:41 ff.).

In the light of the generalization provided by De Broglie one may ask: if it is possible to describe or explain entities qualified by energy in terms of two mutually exclusive experimental perspectives, namely as *particles* and as *waves*, is it then still meaningful to speak about their *unitary structure*? This question puts the finger exactly on that point where the special scientific description

---

1 Cf. Margenau, 1982:187, 196-197.
2 Light quanta are called photons and similar to the neutrino they possess a zero mass.
3 Interference phenomena were established after Michelson – round 1880 – designed an interferometer capable of cutting light and afterwards recombining it. Thus one ends up with the same light beam – with slightly less energy. The remarkable result was that the sum did not produce light but darkness! However, when one of the two halves was blocked with a piece of black paper the other halve did appear. Seemingly the only way to explain what happened here is to assume that the interference of the split light-waves cancel out each other when reunited.

reaches its limits and needs to fall back upon a perspective transcending the confines of special scientific inquiry. What is here required is some or other philosophical account transcending the mere combination of one or more (modally delimited) special scientific points of view. The *idea* of the unity and identity of an entity could never be provided to us by theoretically explicating various modal functions, simply because this underlying unity is presupposed in all theoretical explanations. In a strict and technical sense this idea of an entity in its totality – preceding the analysis of its modal aspects – refers to an individual whole embedded in the inter-modal and inter-structural coherence of reality, to an entity emerged in the dept-layer of an all-embracing *temporality* transcending genuine concept-formation and only to be approximated in a *concept-transcending idea*.

A deepening of this basic (transcendental) idea occurs when – through theoretical reflection and investigation – the dimension of micro-structures is unveiled (the micro-world with atoms and sub-atomic particles). It is important in this context, however, to realize that concepts such as *particle, field*, and *wave* are not *type concepts* but *modal functional concepts* (sometimes referred to as elementary basic concepts of physics). Consequently, the terms particle and wave analogically reflect retrocipatory structural moments within the structure of the kinematical aspect, namely *movement multiplicity* (numerical analogy) and *movement extension* (spatial analogy). These facets are deepened in physically qualified entities and could be approximated in physical theory from the perspective of mathematical anticipations to the physical aspect – compare Shrödinger's wave function formulated in terms of differential equations.

Since number, space and movement remain irreducible aspects regardless of the nature and type of entities functioning within them (their modal universality), it is also from this perspective understandable why the functionally distinct concepts *particle* and *wave* cannot be reduced to each other – a state of affairs supported by experimental data. Irreducible modal perspectives may also serve as modes of scientific explanation.

Born, Pyrmont and Biem reject the struggle with a *dualism* in this context. They hold that it increasingly becomes clear that "nature could neither be described by particles alone, nor solely through waves," because a proper understanding cannot toggle between a "particle image [*Teilchenbild*]" and a "wave image [Wellenbild]." This leads to a unitary view of physical systems.[1] What we have called *modes of explanation* these authors designate as "Darstellungen" (representations) – and they specifically mention three distinct (but present at once) modes of explanation: an "Ortsdarstellung" (spatial position), a "Wellendarstellung" (impulses or velocities – kinematic explanation) and an "Energiedarstellung" (the physical mode of explanation) (Born, Pyrmont, Biem, 1967/68:416-417).

---

1   "Mit der Quantentheorie erfaßt man so alle Systeme einheitlich, ..."

Bibliography

Agassi, J. & Cohen, R.S. (editors) 1982. *Scientific Philosophy Today, Essays in Honour of Mario Bunge*, Boston Studies in the Philosophy of Science, Volume 67, Dordrecht, Boston, London, 1982.
Agazzi, E. 2001. Philosophy of Nature and Natural Sciences. In: *Philosophia Naturalis*, 38(1-2):1-23.
Apolin, A. 1964. Die Geschichte des Ersten und Zweiten Hauptzatzes der Wärmetheorie und ihre Bedeutung für die Biologie. In: *Philosophia Naturalis*:
Aristotle. 2001. *The Basic Works of Aristotle. Edited by Richard McKeon with an Introduction by C.D.C. Reeve.* (Originally published by Random House in 1941). New York: The Modern Library.
Bartle, R.G. 1964. *The Elements of Real Analysis.* New York: John Wiley & Sons, Inc.
Becker, O. 1965. Preface. In: *Zur Geschichte der griechischen Mathematik*, Wege der Forschung, Band 43, Darmstadt.
Berberian, 1994. *A First Course in Real Analysis.* New York: Springer.
Bernays, P. 1976. *Abhandlungen zur Philosophie der Mathematik.* Darmstadt: Wissenschaftliche Buchgesellschaft.
Beth, E.W. 1965. *Mathematical Thought.* New York: D. Reidel Publishing Company.
Bohr, N. 1966. *Atoomtheorie en Natuurbeschrijving*, Aula-uitgawe, Antwerpen 1966.
Born, M., Pyrmont, B. and Biem, W. 1967-1968. Dualismus in der Quantentheorie. In: *Philosophia Naturalis*, Volume 10 (pp.411-418).
Borsche, T. 1976. Individuum, Individualität. In: *Historisches Wörterbuch der Philosophie*, Eds. J. Ritter, K. Gründer & G. Gabriel, Volume 4 (pp.310-323). Basel-Stuttgart: Schwabe & Co.
Breuer, T. 1997. Universell und unvolständig: Theorien über alles? In: *Philosophia Naturalis*. 34:1-20.
Brouwer, L.E.J. 1911. Beweis der Invarianz der Dimensionenzahl. *Mathematische Annalen*, LXX (pp.161-165).
Brouwer, L.E.J. 1924. Bemerkungen zum natürlichen Dimensionsbegriff. In: Brouwer, 1976, (2)554-557.
Brouwer, L.E.J. 1976. *Collected Works*, Volume 2, *Geometry, Analysis, Topology* and *Mechanics*, Editor Hans Freudenthal. Amsterdam: North Holland.
Cantor, G. 1962. *Gesammelte Abhandlungen Mathematischen und Philosophischen Inhalts.* Hildesheim: Oldenburg Verlag (1932).
Cassirer, E. 1911. *Das Erkenntnisproblem in der Philosophie und Wissenschaft der neueren Zeit.* Volume II. Berlin: Verlag Bruno.
Cassirer, E. 1953. *Substance and Function.* First edition of the English translation of Substanzbegriff und Funktionsbegriff: 1923; (First German edition 1910). New York: Dover.
Clouser, R.A. 2005. *The Myth of Religious Neutrality: An Essay on the Hidden Role of Religious Belief in Theories.* Notre Dame: University of Notre Dame Press (new revised edition, first edition 1991).
De Vleeschauwer, H.J. 1952. *Handleiding by die Studie van die Logika en die Kennisleer*, Pretoria: Uitgewery J.J. Moerau & Kie.
Derrida, J. 2002. Force of Law, The "Mystical Foundation of Authority". In: *Acts of Religion. Edited and with an Introduction by Gil Anidjar*, New York: Routledge.

Descartes, R. 1965. *A Discourse on Method, Meditations and Principles*, translated by John Veitch, Introduced by A.D. Lindsay. London: Everyman's Library.

Dooyeweerd, H. 1997. *A New Critique of Theoretical Thought*, Collected Works of Herman Dooyeweerd, A-Series Vols. I-IV, General Editor D.F.M. Strauss. Lewiston: Edwin Mellen.

Dummett, M.A.E. 1978. *Elements of Intuitionism*. Oxford: Clarendon Press.

Dummett, M.A.E. 1995. Frege, *Philosophy of Mathematics*. Second Printing. Cambridge: Harvard University Press.

Ebbinghaus, H.-D. 1977. *Einführung in die Mengenlehre*. Darmstadt: Wissenschaftliche Buchgesellschaft.

Ebbinghaus, H.-D., Hermes, H., Hirzebruch, F., Koecher, M., Mainzer, K., Neukirch, J., Prestel, A. And Remmert, R. 1995. *Numbers* (corrected third printing). New York: Springer.

Einstein, A. 1959. Autobiographical Notes. In: *Albert Einstein, Philosopher-Scientist.* Edited by P.A. Schilpp. New York: Harper Torchbooks.

Einstein, A. 1985. *Relativity, the Special and General Theory*. Bristol: Arrowsmith (reprint of the first 1920 translation).

Felgner, U. (Editor) 1979. *Mengenlehre*. Darmstadt: Wissenschaftiche Buchgesellschaft.

Fern, R.L. 2002. *Nature, God and Humanity*, Cambridge: University Press.

Fischer, L. 1933. *Die Grundlagen der Philosophie und der Mathematik*, Leipzig.

Fraenkel, A. 1928. *Einleitung in die Mengenlehre*. Berlin: Verlag von Julius Springer (3rd impresion).

Fraenkel, A., Bar-Hillel, Y., Levy, A. & Van Dalen, D. 1973. *Foundations of Set Theory*, 2nd revised edition. Amsterdam: North Holland.

Fränkel, H. 1968. Zeno von Elea im Kampf gegen die Idee der Vielheit. In: *Um die Begriffswelt der Vorsokratiker*, Wege der Forschung, Band IX, Editor Hans-Gerog Gadamer, Darmstadt: Wissenschaftliche Buchgesellschaft (pp.425 ff.).

Frege, G. 1884. *Grundlagen der Arithmetik*. Breslau: Verlag M & H. Marcus (Unaltered reprint, 1934).

Frege, G. 1979. *Posthumous Writings*. Oxford: Basil Blackwell.

Gödel, K. 1995. Collected Works, Volume III, *Unpublished essays and lectures*, edited by Solomon Feferman (Editor-in-Chief), Oxford: University Press.

Greene, B. 2003. *The Elegant Universe*. New York: W.W. Norton & Company.

Grelling, K. 1910. *Die Axiome der Arithmetik mit besonderer Berücksichtigung der Beziehungen zur Mengenlehre*, Inaugural-Dissertation. Göttingen: Dieterscischen Universitätsa-Buchdruckerei.

Grünbaum, A. 1952. A Consistent Conception of the Extended Linear Continuum as an Aggregate of Unextended Elements. In: *Philosophy of Science*, Vol.19, nr.2, April (pp.288-306).

Guthrie, W.K.C. 1980. *A History of Greek Philosophy*. Volume II. The Presocratic Tradition from Parmenides to Democritus. Cambridge: Cambridge University Press.

Habermas, J. 1996. *Between facts and norms: contributions to a discourse theory of law and democracy*, translated by William Rehg; 2nd print, Cambridge: Massachusetts: MIT Press.

Habermas, J. 1998. *Faktizität und Geltung. Beiträge zur Diskurstheorie des Rechts und des demokratischen Rechtsstaats*. Frankfurt am Main: Suhrkamp edition (first edition 1992).
Hartmann, N. 1957. *Kleinere Schriften*, Volume II, Berlin: De Gruyter.
Hawking, S.W. 1988. *A Brief History of Time*. London: Transworld Publishers.
Herder, J.G. (Editor Bernhard Suphan). 1877. Herders sämmtliche Werke. Berlin, Weidmann, (1877-1913).
Herold, N. 1974. Der Gesetzesbegriff in Philosophie und Wissenschaftstheorie der Neuzeit. In: *Historisches Wörterbuch der Philosophie*, Eds. J. Ritter, K. Gründer & G. Gabriel, Volume 4 (pp.501-514). Basel-Stuttgart: Schwabe & Co.
Heyting, A. 1971. *Intuitionism. An Introduction*. Amsterdam: North Holland Publishing Company.
Heyting, A.T.A. 1949. *Spanningen in de Wiksunde*. Groningen & Batavia: P. Noordhoff.
Hilbert, D. 1925. Über das Unendliche, *Mathematische Annalen*, Vol.95, 1925: 161-190.
Hilbert, D. 1970. *Gesammelte Abhandlungen*, Vol.3, Second Edition, Berlin: Verlag Springer.
Hucklenbroich, P. 1980. Der physikalische Begriff der Materie. In: Ritter, *Historisches Wörterbuch der Philosophie*. Volume 5, Stuttgart: Schwabe & Co Verlag (pp.921-924).
Hurewicz, W. and Wallman, H. 1959: *Dimension Theory*, 5th edition. Princeton: University Press.
Janich, P. 1975. Tragheitsgesetz und Inertialsysteem. In: *Frege und die moderne Grundlagenforschung*, red. Chr. Thiel, Meisenheim am Glan: Hain.
Kant, I. 1787. *Kritik der reinen Vernunft*, 2nd Edition (references to CPR B). Hamburg: Felix Meiner edition (first edition 1781).
Kiontke, Siegfried 2006. *Physik biologischer Systeme, Die erstaunliche Vernachlässigung der Biophysik in der Medizin*. München: Mintzel.
Klein, F. 1925. *Elementar Mathematik vom höheren Standpunkte aus, Geometrie*. Third Edition, Volume II. Belrin: Verlag Julius Springer.
Klein, F. 1932. *Elementary Mathematics from an Advanced Standpoint*. London: Macmillan.
Klein, F. 1939. *Elementary Mathematics from an Advanced Standpoint. Geometry*. London: Dover Publications.
Kline, M. 1980. *Mathematics, The Loss of Certainty*. New York: Oxford University Press.
Kobusch, Th. 1976. Individuum, Individualität. In: *Historisches Wörterbuch der Philosophie*, Eds. J. Ritter, K. Gründer & G. Gabriel, Volume 4 (pp.299-304). Basel-Stuttgart: Schwabe & Co.
Koestler, A. and Smythies, J.R. (Eds.) 1972. *Beyond Reductionism*. New York: Macmillan.
Lakoff, G. & Núñez, R.E. 2000. *Where Mathematics Comes From, How the Embodied Mind Brings Mathematics into Being*. New York: Basic Books.
Laugwitz, D. 1986. *Zahlen und Kontinuum. Eine Einführung in die Infinitesimalmathematik*. Mannheim: B.I.-Wissenschaftsverlag.
Leibniz, G.W.L. 1976. *Philosophical Papers*. Edited by Leroy E. Loemker. Synthese Historical Library, Volume 2. Dordrecht-Holland: D. Reidel (first print 1969).

Lorenzen, P. 1952. Ueber die Widerspruchfreiheit des Unendlichkeitsbeg riffes. In: *Studium Generale, Zeitschrift für die Einheit der Wissenschaft en im Zusammenhang ihrer Begriffsbildungen und Forschungsmethoden*, Vol. 10, Berlin.
Lorenzen, P. 1960. *Die Entstehung der exakten Wissenschaften*. Berlyn: Springer-Verlag.
Lorenzen, P. 1968. Das Aktual-Unendliche in der Mathematik. In: *Methodisches Denken*. Frankfurt am Main: Suhrkamp Taschenbuch Wissenschaft (73), pp.94-119 – also published in Meschkowski, 1972:157-165.
Lorenzen, P. 1972. Das Aktual-Unendliche in der Mathematik, In: Meschkowski, 1972:157-165.
Mac Lane, S. 1986. *Mathematics: Form and Function*. New York: Springer-Verlag,
Maddy, P. 2005. Three forms of Naturalism. Chapter 13 (pp.437-459). In: Shapiro, 2005.
Maier, A. 1949. *Die Vorläufer Galileis im 14. Jahrhundert*. Roma: Edizioni di Storia e letteratura.
Maier, A. 1964. *Ausgehendes Mittelalter*. Vol.I, Rome.
Margenau, H. 1982. *Physics and the Doctrine of Reductionism*. In: Agassi, & Cohen, 1982.
Meschkowski, H. (Editor) 1972. *Grundlagen der modernen Mathematik*, Darmstadt: Wissenschaftliche Buchgesellschaft.
Monk, J.D. 1970. On the Foundations of Set Theory, in: *The American Mathematical Monthly*, Vol.77.
Myhill, J. 1972. What is a real number? In: *American Mathematical Monthly*, 79:748-754.
Oeing-Hanoff, L. 1976. Individuum, Individualität – Hoch- und Spätskolastik, in: *Historisches Wörterbuch der Philosophie*, Eds. J. Ritter, K. Gründer & G. Gabriel, Volume 4 (pp. 304-310). Basel-Stuttgart: Schwabe & Co.
Planck, M. 1973. *Vorträge und Erinnerungen*, 9th reprint of the 5th impression. Darmstadt: Wissenschaftliche Buchgesellschaft.
Popper, K.R. Scientific Reduction and the Essential Incompleteness of All Science. In: Dobzhansky, 1974 (pp. 259-284).
Posy, C. 2005. Intuitionism and Philosophy. In: Shapiro, 2005:319-355.
Quine, W.V.O. 1953. *From a Logical Point of View*. Cambridge Massachusetts: Harvard University Press.
Resnik, M.D. 1997. *Mathematics as a science of patterns*. Oxford: Clarendon Press.
Robinson, A. 1966. *Non-standard analysis*. Amsterdam: North-Holland (second edition, 1974).
Rollwagen, W. 1962. *Das Elektron der Physiker*. Munich: Max Hüber.
Russell, B. 1897. *An essay on the foundations of geometry*. Cambridge: University Press.
Savage, C.W. and Ehrlich, P. (Eds.) 1992. *Philosophical and Foundational Issues in Measurement Theory*. New Yersey: Lawrence Erlbaum Associates, Inc.
Shapiro, S. 2005 (Editor). *The Oxford Handbook of Philosophy of Mathematics and Logic*. Oxford: Oxford University Press.
Sikkema, A. 2005. A Physicist's Reformed Critique of Nonreductive Physicalism and Emergence, *Pro Rege*, 33:4 (June):20-32.

Smith, G.L., 1994. *On Reductionism*. Sewanee, Tennessee – available on the WEB at: http://smith2.sewanee.edu/texts/Ecology/OnReductionism.html (accessed on 28-03-2008).
Stafleu, M.D. 1987. *Theories at Work*: On the Structure and Functioning of Theories in Science, in Particular during the Copernican Revolution, Lanham: University Press of America.
Stafleu, M.D. 1989. *De Verborgen Structuur*. Amsterdam: Buijten & Schipperheijn.
Stafleu, M.D. 2002. *Een Wereld vol Relaties*. Amsterdam: Buijten & Schipperheijn.
Steffens, H.J. 1979. *James Prescott Joule and the concept of energy*. Folkstone, Eng.: Dawson, New York: Science History Publications.
Stegmüller W. 1977. *Collected Papers on Epistemology, Philosophy of Science and History of Philosophy*. Volumes I and II. Dordrecht-Boston: D. Reidel Publishing Company.
Stegmüller W. 1987. *Hauptströmungen der Gegenwartsphilosophie*. Volume III, Stuttgart: Alfred Kröner Verlag.
Stegmüller, W. 1969. *Main Currents in Contemporary German, British and American Philosophy*. Dordrecht: D. Reidel Publishing Company, Holland.
Strauss, D.F.M. 1982. The Place and Meaning of Kant's Critique of Pure Reason (1781) in the legacy of Western philosophy. In: *South African Journal of Philosophy*, Volume 1, (pp.131-147).
Strauss, D.F.M. 2000. Kant and modern physics. The synthetic a priori and the distinction between modal function and entity. In: *South African Journal of Philosophy*, 2000, pp.26-40.
Strauss, D.F.M. 2001. *Paradigms in Mathematics, Physics, and Biology – their Philosophical Roots*. Bloemfontein: Tekskor (Revised Edition, 2004).
Strauss, D.F.M. 2008. *Philosophy: Discipline of the Disciplines*. (To Appear).
Tait, W. 2005. *The Provenance of Pure Reason, Essays in the Philosophy of Mathematics and Its History*. Oxford: University Press.
Vaihinger, H. 1922. *Die Philosophie des Als Ob: System der theoretischen, praktischen und religiösen Fiktionen der Menschheit auf Grund eines idealistischen Positivismus, mit einem Anhang über Kant und Nietzsche*, 7th and 8th Edition. Leipzig: Meiner.
Van Fraassen, Bas C. 1991. *Laws and symmetry*. Oxford: Clarendon Press.
Van Melsen, A.G.M. 1975. Atomism. In: *Encyclopedia Britannica*, 15th edition, London, Volume 2, pp.346-351.
Vogel, H. 1961. *Zum Philosophischen Wirken Max Plancks*. Seine Kritik am Positivismus. Berlin: Akademie-Verlag.
Von Fritz, K. 1965. Die Entdeckung der Inkommensurabilität durch Hippasos von Metapont. In: Becker, 1965:271-307.
Von Weiszäcker, C.F. 1993. *Der Mensch in seiner Geschichte*. München: DTV.
Von Weizsäcker, C.F. 2002. *Große Physiker, Von Aristoteles bis Werner Heisenberg*. München: Deutscher Taschenbuch Verlag.
Wang, H. 1988: *Reflections on Gödel*. Cambridge Massachusetts: MIT Press.
White, M.J. 1988. On Continuity: Aristotle versus Topology? In: *History and Philosophy of Logic*, 9:1-12.
Wolff, K. 1971. Zur Problematik der absoluten Überabzählbarkeit, in: *Philosophia Naturalis*, Band 13, 1971.

Index of Subjects

A
a priori 6, 35, 54
absolute rigour 5
acceleration 39, 43
actual infinite 21, 24-28, 30, 65
ad infinitum 29
algebraic number 63
analogical basic concepts 55
analogies 10, 17-18, 23, 50, 54, 59
analogy of number 17
anticipation to a retrocipation 26
anticipations 22-23, 50
anticipatory coherence 28
antinomies 1, 5, 36, 58
arithmetical
 – aspect 18, 22-23, 52
 – language 26
 – laws 8-9, 16
arithmeticism 6, 11, 17, 31, 37
arithmetization 15, 23-24, 31
astronomy 46
astrophysics 46
at once infinite 21, 26, 28, 30-32, 65
atomic number 61
atomistic conception 47
axiomatic formalism 5
axiomatization 21

B
basic concepts 15, 23, 54-55, 57
being distinct 50, 52
biotic heterogeneity 49

C
causing force 39
chemical bond 60-61
classical
 – mathematics 3
classical mechanics 39
complete arithmetization of mathematics 23-24
compound basic concepts 15, 54-55, 57
concept
 – of a natural law 57-58
 – of law 9
 – of number 14, 17, 27
concept-transcending knowledge 44, 51
conceptual knowledge 50-51
concrete material extension 37
constant quantity 28
constructive interpretation 25
continuity conception 47
continuous
 – becoming 49
 – change 49
 – quantity 13, 15
corpuscular 41
cosmology 46
countability 40, 64

D
deceleration 39
degenerate intervals 30
determinism 33
differentia specifica 13
dimension of rationality 34
discrete
 – orbits 58
 – quantity 13, 15, 22, 52
discreteness and continuity 4, 15
distance axioms 30
domain
 – of number 15-17
 – of space 12, 16, 18

E
electromagnetic radiation 58
electrons 41, 58-59, 61
electron-shells 58
elementary
 – basic concepts 23, 54
 – particles 58-59
empty space 35-36
energy
 – conservation 44
 – constancy 44, 46
 – operation 22, 43, 51-52
enkaptic totality 61
Enlightenment 19
entitary analogies 18
eternity 50
Euclidean geometry 10
everything
 – coheres 45, 53
 – is unique 53
 – remains identical 53
extended linear continuum 21, 29

F

factual
 - extension 17, 20, 29
 - side 9-10, 16-20, 22-23, 25-26, 31, 57
faith in reason 33
fermions 58
field axioms 8
formalism 5
formalization 3
foundational
 - crisis 5
 - function 59-60
functional modes 22

G

geno-types 61
genus proximum 13, 15
geometrical
 - figures 4, 10, 53
 - intuition 14, 23
 - sum 10
geometrization 5-6, 10
given at once 24, 26
gravitation 43
gravitational force 43

H

historicism 1-2

I

idea-knowledge 44, 52
impetus 38-39
incommensurability 4, 6
incommunicable 49
indestructibility of matter 47
indeterminism 33
individuality 44, 48-53, 62
individuality-structures 62
induction 23
inertia 38-40
infinite
 - decimal fraction 24, 64
 - divisibility 26, 36-38, 46-47
 - force 43
 - multiplicity 27-28
 - succession 27
 - totality 27-28
infinitely divisible 19, 29-30, 36, 38, 60
infinitesimals 28
infinitum
 - simultaneum 28
 - successivum 28

infinity 14, 24, 26-28, 33, 39, 65
intuition 3-5, 9, 14-15, 23, 25-26, 28, 37, 50-52, 64
intuitionists 3
invariance 20
irrational numbers 6, 11, 34-35
irreducibility 6, 13, 18, 21, 24, 28, 30-32
irreversible processes 44

K

kinematic persistence 48

L

law of inertia 38-40
law-conformative 9, 56, 65
law-order 58
law-side 9-10, 12, 16-19, 23-24, 26, 31, 44, 57
law-spheres 21
Law-Word 1
line segments 10, 30
linear Cantorean continuum 29
line-stretch 19-20
logicism 5
logicist 3, 5-6

M

material things 37, 44-46, 48, 61
mathematical space 46-47, 60
matter and energy 33
meaning-nucleus 18, 22, 31
measure-theory 30
metaphors 17-18
modal
 - analogies 50
modal aspects 18, 21, 25, 44-45, 48, 54-57, 60, 62
 - functions 17-18, 54
 - laws 18, 45, 56-58
 - terms 44, 52-53
 - universality 22, 43-46
mode of being 18
modes of explanation 18, 36, 39-42, 44-48, 51-53
multiplicity 11, 21, 23, 25-28, 36-37, 57
mystery of matter 44

N

natural
 - laws 53-54
 - numbers 8-9, 17, 23, 26, 63, 65
necessity 41, 44, 54
neo-Platonism 49
neutrality postulate 1

neutrons 41, 59
non-denumerability 25, 65
non-numerical ratios 10
non-reductionist 1-2, 6, 21, 33, 62
non-reductionist ontology 1, 6, 33, 62
number of dimensions 14
numerals 7, 21
numerical
– analogies 18, 50, 59
– aspect 22-23, 26-27, 37
– difference 18
– relationships 4, 34
– time-order of succession 23, 26, 30

O
objective and neutral 1, 4, 6
objective reality 21-22
ontic order 54
operation of addition 7
ordinal number 23

P
partial perspective 44, 46
periodic table 61
permanence 38, 54
persistence 39, 48
philosophy of nature 33
phoronomy 52
physical
– aspect 33, 42-46, 48, 52, 54-55, 58
– changefulness 48
– entities 57-58, 60
– force 39
– homogeneity 49
– quantity 42
– space 46-47, 60
polinomial equation 63
positivism 1-2
postmodernism 1-2
potential infinite 24, 65
primitive terms 13, 18, 21, 23
principle of individualization 48
protons 41, 59, 61

Q
quarks 59
quasi-spatial 26

R
radio-activity 43
rational
– knowledge 34
– numbers 6, 9, 11, 26, 30, 34-35, 63

real
– analysis 12
– numbers 8, 21, 24-25, 28-32, 34, 64-65
rectilinear motion 39
reductionism 1-2, 62
regular pentagram 34
res extensa 37
retrocipations 22-23, 50

S
semi-disclosed 26
semiperceptions 22
sensory objects 35
set theory 3, 5, 7, 9, 15-16, 25-26, 29, 32, 37, 52
similarities and differences 17
simultaneity 21, 23, 26-28, 30-31
space metaphysics 5, 37, 45
spatial
– addition 6-7
– aspect 10, 17-20, 25-27, 29, 31, 35-37, 51-52, 56, 60
– magnitudes 10
– object 20-21
– subject 11-12, 16, 20-21, 25, 29, 37
– terms 23
– time-order of simultaneity 30
special sciences 1, 27, 33
static forms 35
straight line 4, 12, 14, 17, 30, 35
structural features 4, 18
subject-object relations 19, 23
subject-subject relations 18, 22-23
super string theory 48
system
– of integers 8
– of rational numbers 9, 30

T
theoretical idea 1
thought
– and being 35
– experiment 39
thought-form 51
time-order
– of simultaneity 27, 30
time-order of succession 23, 26, 30
topologically invariant 20
topology 31
totality 23-28, 30-31, 37, 44, 52, 59, 61
totality view 44
transcendental-empirical method 57
transformability of matter 47

type-laws 56-57
typicality 44-45

U

ultimate commitment 1, 33-34
uncompleted infinity 27, 65
undefined term 12, 26
uniform flow 52
uniformity 56-57
universal theory 45

universality and individuality 50-51
universally valid 58, 62

V

variability types 61

W

wholeness 23, 25-26, 36-37, 48-49
whole-parts relation 23, 30-31, 36-37, 49-50, 60
Wirkungsquantum 42

Index of Names

A
Agassi 68, 71
Agazzi 33, 68
Apolin 44, 68
Aristotle 13, 15, 19, 48-49, 52, 68, 72

B
Bar-Hillel 69
Bartle 8, 68
Becker 35, 65, 68, 72
Berberian 8, 12, 68
Bernays 14, 16-17, 21, 23-27, 31, 68
Beth 3, 68
Biem 67, 68
Bohr 41, 43, 58, 66, 68
Born 67, 68
Borsche 49, 68
Breuer 68
Brouwer 64, 68

C
Cantor 5, 10, 14, 16-17, 20, 24-25, 28-31, 63-65, 68
Cassirer 22, 53-54, 68
Clouser 68
Cohen 68, 71

D
De Broglie 66
De Vleeschauwer 51, 68
Derrida 55, 68
Descartes 5, 10, 37-38, 40-41, 43, 68
Dooyeweerd 22, 28, 50, 57, 60, 62, 69
Dummett 5, 21, 27, 69

E
Ebbinghaus 8-9, 69
Ehrlich 10, 71
Einstein 40, 42-44, 48, 66, 69
Eisberg 66

F
Felgner 16, 69
Fern 3-4, 69
Fischer 27, 69
Fraenkel 5, 9, 15, 25, 37, 52, 65, 69
Fränkel 36, 69
Frege 5-6, 11, 14-16, 21, 40, 69-70

G
Gödel 5, 21-22, 26, 69, 72
Greene 48, 69
Grelling 16, 69
Grünbaum 21, 29-30, 69
Guthrie 36, 69

H
Habermas 55, 69
Hartmann 51, 69
Hawking 41, 70
Herder 49, 70
Herold 53, 70
Heyting 3, 5, 25, 70
Hilbert 5, 12-13, 18, 21, 24, 41, 70
Hucklenbroich 70
Hurewicz 19-20, 70

J
Janich 42-43, 70

K
Kant 5, 19, 22, 37, 51, 54, 70, 72
Kiontke 59, 70
Klein 7, 9, 47, 69-70
Kline 3, 70
Kobusch 70
Koestler 70

L
Lakoff 41, 70
Laugwitz 4-5, 14, 31, 47, 70
Leibniz 40-41, 49, 70
Levy 69
Lorenzen 24-25, 27, 35, 42, 70-71

M
Mac Lane 4, 12-13, 71
Maddy 16, 71
Maier 28, 38-40, 42, 71
Margenau 66, 71
Meschkowski 63, 71
Monk 3, 71
Myhill 9, 71

N
Núñez 41, 70

O
Oeing-Hanoff 49, 71

P
Planck 42-44, 71-72
Popper 1, 71
Posy 21, 71
Pyrmont 67, 68

Q
Quine 2, 9, 71

R
Resnik 15, 71
Robinson 28, 71
Rollwagen 48, 60, 71
Russell 4-5, 10, 12-13, 42, 71

S
Savage 10, 71
Shapiro 71
Sikkema 22, 71

Smith 2, 71
Stafleu 41, 43, 57, 60, 62, 71-72
Steffens 44, 46, 72
Stegmüller 3-4, 33, 46-47, 57, 72
Strauss 51-52, 65, 67, 69, 71-72

T
Tait 14, 72

V
Vaihinger 27, 72
Van Dalen 69
Van Fraassen 56, 72
Van Melsen 62, 72
Vogel 41, 72
Von Fritz 10, 72
Von Weiszäcker 50, 72

W
Wallman 19-20, 70
Wang 21-22, 26, 72
White 4, 72
Wolff 25, 72

www.ingramcontent.com/pod-product-compliance
Lightning Source LLC
Chambersburg PA
CBHW032048290426
44110CB00012B/998